A Level
Chemistry
for OCR
Year 2

A

Rob Ritchie

OXFORD
UNIVERSITY PRESS

OXFORD
UNIVERSITY PRESS

Great Clarendon Street, Oxford, OX2 6DP, United Kingdom

Oxford University Press is a department of the University of Oxford. It furthers the University's objective of excellence in research, scholarship, and education by publishing worldwide. Oxford is a registered trade mark of Oxford University Press in the UK and in certain other countries

First published in 2017

British Library Cataloguing in Publication Data
Data available

978-019-835777-3

10 9 8 7

Paper used in the production of this book is a natural, recyclable product made from wood grown in sustainable forests. The manufacturing process conforms to the environmental regulations of the country of origin.

Printed in Great Britain

Acknowledgements

Cover: Kusska/Shutterstock

Artwork by Q2A Media

AS/A Level course structure

This book has been written to support students studying for OCR A Level Chemistry A. It covers the Year 2 modules from the specification. The modules covered are shown in the contents list, which also shows you the page numbers for the main topics within each module.

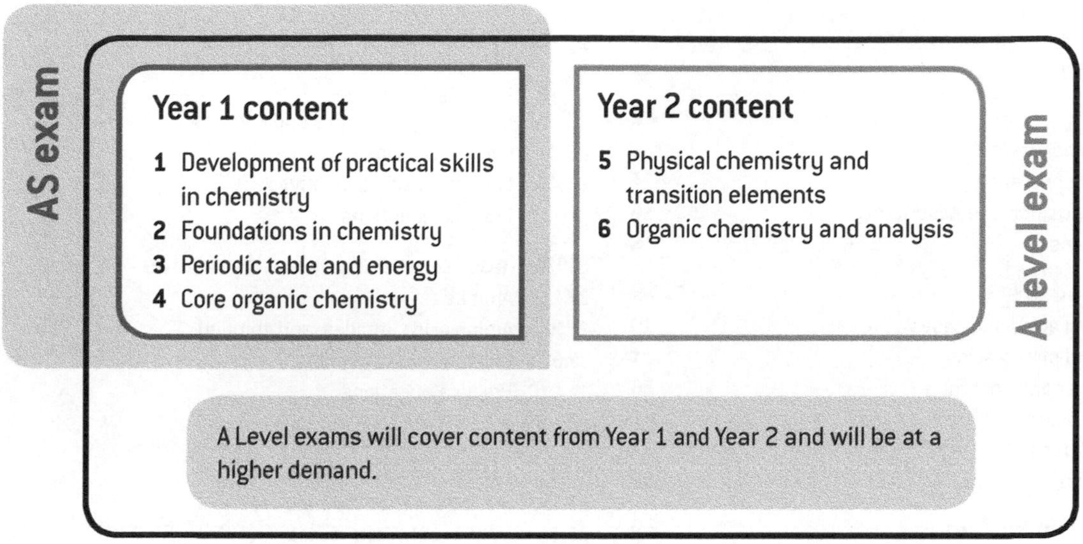

Contents

How to use this book

This book contains many different features. Each feature is designed to support and develop the skills you will need for your examinations, as well as foster and stimulate your interest in chemistry.

Worked example

Step-by-step worked solutions.

Summary Questions

1 These are short questions at the end of each topic.

2 They test your understanding of the topic and allow you to apply the knowledge and skills you have acquired.

3 The questions are ramped in order of difficulty. Lower-demand questions have a paler background, with the higher-demand questions having a darker background. Try to attempt every question you can, to help you achieve your best in the exams.

Specification references

- At the beginning of each topic, there are specification references to allow you to monitor your progress

Revision tip

Revision tips contain prompts to help you with your understanding and revision.

Synoptic link

These highlight the key areas where topics relate to each other. As you go through your course, knowing how to link different areas of chemistry together becomes increasingly important. Many exam questions, particularly at A Level, will require you to bring together your knowledge from different areas.

Practice questions at the end of each chapter, including questions that cover practical and math skills.

Chapter 19 Practice questions

1 What is the expression for K_c for the equilibrium:
$C(s) + H_2O(g) \rightleftharpoons CO(g) + H_2(g)$?

A $\frac{[CO(g)][H_2(g)]}{[C(s)][H_2O(g)]}$ B $\frac{[C(s)][H_2O(g)]}{[CO(g)][H_2(g)]}$

C $\frac{[CO(g)][H_2(g)]}{[H_2O(g)]}$ D $\frac{[H_2O(g)]}{[CO(g)][H_2(g)]}$

(1 mark)

2 0.300 mol of HBr(g) is added to a 1.00 dm^3 container and left to reach equilibrium at constant temperature: $2HBr(g) \rightleftharpoons H_2(g) + Br_2(g)$.
At equilibrium, 0.0500 mol of both $H_2(g)$ and $Br_2(g)$ are present.
What is the value of K_c?

A 0.0125 **B** 0.0278 **C** 0.0625 **D** 0.250 *(1 mark)*

3 N_2 reacts with H_2 in the equilibrium:

$N_2(g) + 3H_2(g) \rightleftharpoons 2NH_3(g)$ $\Delta H = -92\,kJ\,mol^{-1}$

The temperature is increased. What is the effect on the equilibrium amount of NH_3 and the value of K_c?

	amount of NH_3	value of K_c
A	decrease	increase
B	increase	decrease
C	increase	increase
D	decrease	decrease

(1 mark)

4 SO_2 reacts with O_2 in the equilibrium:

$2SO_2(g) + O_2(g) \rightleftharpoons 2SO_3(g)$ $\Delta H = -197\,kJ\,mol^{-1}$

The pressure is decreased. What is the effect on the equilibrium position and the value of K_c?

	equilibrium position	value of K_c
A	shifts to right	increase
B	shifts to left	decrease
C	shifts to right	no change
D	shifts to left	no change

(1 mark)

5 CO reacts with H_2 to form the equilibrium: $CO(g) + 2H_2(g) \rightleftharpoons CH_3OH(g)$
0.100 mol of CO(g) and 0.200 mol H_2 are mixed in a 500 cm^3 container.
At equilibrium, 80% of CO has reacted.

a Determine the equilibrium concentrations. *(3 marks)*

b Write the expression for K_c. *(1 mark)*

c Calculate K_c, to three significant figures. Include units. *(2 marks)*

d The pressure is doubled.
Explain the effect on the equilibrium position in terms of K_c. *(3 marks)*

6 SO_2 reacts with O_2 to form the equilibrium:

$2SO_2(g) + O_2(g) \rightleftharpoons 2SO_3(g)$ $\Delta H = -198.2\,kJ\,mol^{-1}$

The equilibrium mixture contains 0.15 mol SO_2, 0.45 mol O_2 and 0.90 mol SO_3.
The total pressure is 200 kPa.

a Determine the equilibrium partial pressures. *(3 marks)*

b Write the expression for K_p. *(1 mark)*

c Calculate K_p. Include units. *(2 marks)*

d The temperature is increased.
Explain the effect on the equilibrium position in terms of K_p. *(3 marks)*

18

18.1 Orders, rate equations, and rate constants

Specification reference: 5.1.1

> **Revision tip**
> For a reactant **A**: [**A**] is shorthand for 'concentration of **A**'.
> $[\mathbf{A}] = \dfrac{\text{amount of } \mathbf{A} \text{ (in mol)}}{\text{volume (in dm}^3\text{)}}$.
> Units of concentration = mol dm^{-3}.

Important terms

Rate of reaction

Rate of reaction is the change in concentration of a reactant or a product in a given time.

$$rate = \frac{\text{change in concentration}}{\text{change in time}}$$

Units

Concentration:	units = mol dm^{-3}
Time:	units in a convenient measurement, often in seconds, s
Rate:	units = $\text{mol dm}^{-3}\,\text{s}^{-1}$ (when time is measured in seconds)

Orders

Order shows how rate is affected by concentration.

Rate is proportional to the concentration raised to the power of the order.

- For order n: rate $\propto [\mathbf{A}]^n$

Zero order

rate $\propto [\mathbf{A}]^0$

- Rate is **not** affected by concentration.

First order

rate $\propto [\mathbf{A}]^1$

- Rate changes by the same factor as a concentration change raised by the power 1.
- If [**A**] is doubled (× 2), rate increases by $2^1 = 2$.

Second order

rate $\propto [\mathbf{A}]^2$

- Rate changes by the same factor as a concentration change raised to the power 2.
- If [**A**] is doubled (× 2), rate increases by $2^2 = 4$

> **Revision tip**
> For order = 0,
> - change concentration: rate does **not** change.
>
> For order = 1:
> - double concentration: rate doubles
> - triple concentration: rate triples
>
> For order = 2:
> - double concentration: rate quadruples (× 4)
> - triple concentration: rate increases by 9 times

Overall order

Overall order is the sum of the individual orders of the reactants.

For two reactants **A** and **B** with orders *m* and *n* respectively.

- Overall order = $m + n$

> **Revision tip**
> If order of **A** = 1 and order of **B** = 2,
> overall order = 1 + 2 = 3

The rate equation

The rate equation shows the link between rate, concentrations, and orders.

The rate constant *k* is the proportionality constant in the rate equation.

The rate equation

For a reaction: **A** + **B** → **C** with orders *m* for **A** and *n* for **B**, the rate equation is given by:

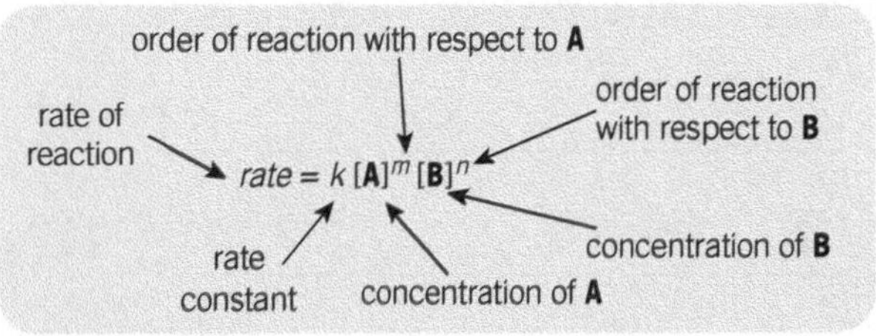

Units of rate constant, *k*

To work out the units of *k*:

- rearrange the rate equation to make *k* the subject
- substitute units into the rearranged rate equation
- cancel common units and write the final units on a single line.

> **Revision tip**
> The units for the rate constant *k* depend upon the number of concentration terms and their orders in the rate equation.

Overall order: 0 $rate = k[A]^0 = k$ units of k = $\mathbf{mol\ dm^{-3}\ s^{-1}}$	**Overall order: 1** $rate = k[A] \quad \therefore k = \frac{rate}{[A]}$ Units of $k = \frac{(\cancel{mol\ dm^{-3}}\ s^{-1})}{(\cancel{mol\ dm^{-3}})} = \mathbf{s^{-1}}$
Overall order: 2 $rate = k[A]^2 \quad \therefore k = \frac{rate}{[A]^2}$ Units of $k = \frac{(\cancel{mol\ dm^{-3}}\ s^{-1})}{(mol\ dm^{-3})^{\cancel{2}}}$ $= \mathbf{dm^3\ mol^{-1}\ s^{-1}}$	**Overall order: 3** $rate = k[A]^2[B] \quad \therefore k = \frac{rate}{[A]^2[B]}$ Units of $k = \frac{(\cancel{mol\ dm^{-3}}\ s^{-1})}{(mol\ dm^{-3})^2\ (\cancel{mol\ dm^{-3}})}$ $= \mathbf{dm^6\ mol^{-2}\ s^{-1}}$

Orders and rate constants from experimental results

Orders from the initial rates method

The initial rate is the rate at the start of a reaction when $t = 0$.

For a reaction, **A** + **B** → **C**, carry out a series of experiments using different initial concentrations of the reactants **A** and **B**.

You must only change one variable at a time:

- Change **[A]** and keep **[B]** constant.
- Change **[B]** and keep **[A]** constant.

> **Revision tip**
> In experiments, initial rates are often determined using 'clock reactions'.
> - In a clock reaction, you measure the time, *t*, until there is a visible change, (usually a colour change).
> - The initial rate is proportional to $\frac{1}{t}$.

Revision tip

Remember that the initial rate is the rate at the start of the experiment when $t = 0$.

Worked example: Determination of a rate constant, *k*

A reacts with **B** as shown below.

$$A + B \rightarrow C$$

The rate of the reaction is investigated and the following experimental results are obtained.

Experiment	[A]/mol dm^{-3}	[B]/mol dm^{-3}	Initial rate/mol dm^{-3} s^{-1}
1	1.00×10^{-3}	2.00×10^{-3}	1.26×10^{-7}
2	2.00×10^{-3}	2.00×10^{-3}	2.52×10^{-7}
3	1.00×10^{-3}	6.00×10^{-3}	1.134×10^{-6}

1 Determine the orders and rate equation.

Step 1: Compare Experiments **1** and **2**,

- [B] stays the same [A] doubles (×2) rate doubles (×2)
- ∴ 1st order with respect to **A**

Step 2: Compare Experiments **1** and **3**,

- [A] stays the same [B] triples (×3) rate increases by ×9
- ∴ 2nd order with respect to **B**

Step 3: The rate equation is $rate = k[A][B]^2$

2 Calculate the rate constant, including units.

Step 1: Rearrange the rate equation, $k = \dfrac{rate}{[A][B]^2}$

Step 2: Substituting values from Experiment 1, $k = \dfrac{1.26 \times 10^{-7}}{(1.00 \times 10^{-3})(2.00 \times 10^{-3})^2} = 31.5$

Step 3: Substituting units into k expression, $\dfrac{(\text{mol dm}^{-3}\ \text{s}^{-1})}{(\text{mol dm}^{-3})\ (\text{mol dm}^{-3})^2} = \text{dm}^6\ \text{mol}^{-1}\ \text{s}^{-1}$

Step 4: Rate constant, $k = 31.5\ \text{dm}^6\ \text{mol}^{-2}\ \text{s}^{-1}$

Revision tip

In the worked example, the results from Experiment 1 were used to calculate k. The results from Experiments 2 and 3 could have been used. Try doing this yourself.

Summary questions

1 A reaction is second order with respect to **A,** first order with respect to **B**, and zero order with respect to **C**.

a What is the overall order of this reaction? *(1 mark)*

b Write the rate equation for this reaction. *(1 mark)*

2 A reaction between **A, B,** and **C** has the rate equation: $rate = k[B]^2[C]$.

a What factor will the rate increase by when:

i the concentration of **A only** is doubled? *(1 mark)*

ii the concentration of **B only** is tripled? *(1 mark)*

iii the concentrations of **A, B,** and **C** are all increased by 4 times? *(1 mark)*

b What are the units of the rate constant of this reaction? *(1 mark)*

3 The results of a series of experiments are shown below.

Experiment	[A]/mol dm^{-3}	[B)]/mol dm^{-3}	Initial rate/mol dm^{-3} s^{-1}
1	0.0250	0.0200	1.25×10^{-4}
2	0.0500	0.0200	2.50×10^{-4}
3	0.0750	0.0400	1.50×10^{-3}

a Determine the rate equation and calculate k, including units. *(3 marks)*

b What is the rate when $[A] = 0.0145$ mol dm^{-3} and $[B] = 0.0256$ mol dm^{-3} *(1 mark)*

18.2 Concentration–time graphs

Specification reference: 5.1.1

Concentration–time graphs

Concentration–time graphs can be plotted from continuous measurements taken during the reaction (continuous monitoring).

Half-life

The half-life of a reactant is the time for its concentration to decrease by half.

- A first order reaction has a constant half-life.

Orders from shapes

Figure 1 shows the shapes of concentration–time graphs for zero order, and first order.

Reaction rates from gradients

To find the rate at any time:

- draw a tangent to the curve at that time
- measure the gradient of the tangent.

The gradient gives the rate of reaction at that time.

Half-life and first order reactions

Monitoring rates by colorimetry

The equation for the reaction between bromine and methanoic acid is shown below.

$$Br_2(aq) + HCOOH(aq) \rightarrow 2Br^-(aq) + 2H^+(aq) + CO_2(g)$$

During the reaction, the concentration of bromine can be monitored continuously using a colorimeter.

The concentration of the other reactant, HCOOH, is kept constant by using a large excess of HCOOH.

- Br_2 reacts to form colourless Br^- ions.
- The orange Br_2 colour disappears.

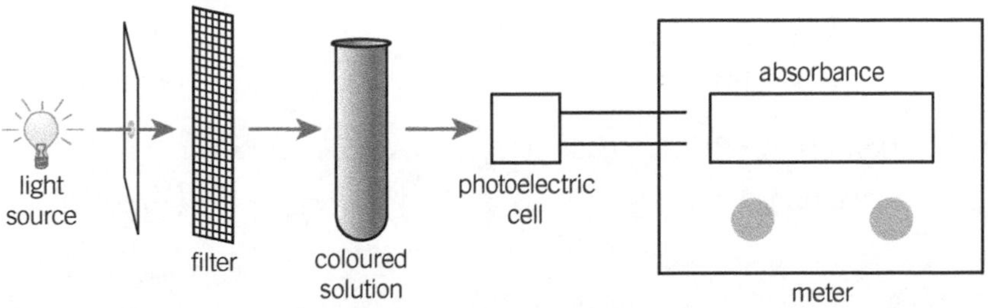

▲ **Figure 2** *A simple colorimeter*

A colorimeter measures the intensity of light passing through a sample. The filter is chosen to be the complementary colour to the colour being absorbed in the reaction. The absorbance reading from the colorimeter is recorded. Absorbance is directly linked to the concentration of the solution.

Synoptic link

Revisit Topic 10.1, Reaction rates, to revise continuous monitoring using gases and for details of measuring reaction rates from gradients.

Zero order

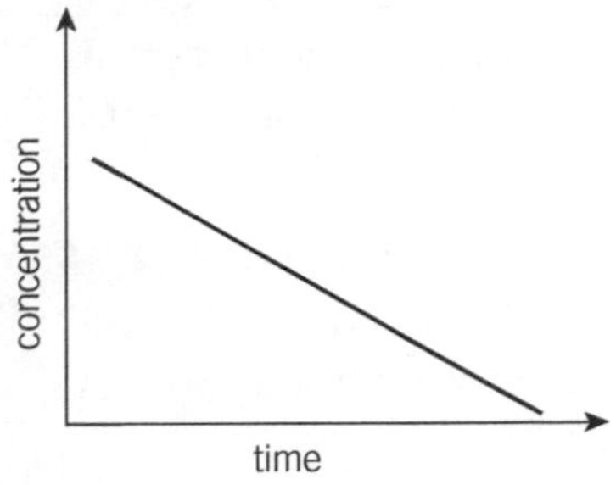

Concentration decreases at a constant rate with time.

First order

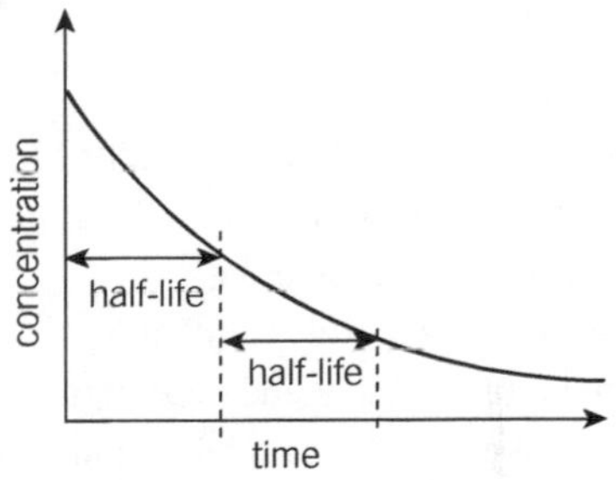

The concentration halves in equal time intervals:

- **constant** half-life.

▲ **Figure 1** *Concentration–time graphs for zero order, and first order*

Revision tip

For a zero order reaction, the rate is equal to the gradient of the straight line of the concentration–time graph.

Synoptic link

Continuous monitoring can also measure the volume of gas or mass loss over time.

For details, see Topic 10.1, Reaction rates.

Figure 3 shows a concentration–time graph for the reaction of Br_2 with HCOOH.

- The half-life is constant at 200 s.
- The reaction is first order with respect to Br_2(aq).
- The rate equation is: rate = $k[Br_2]$

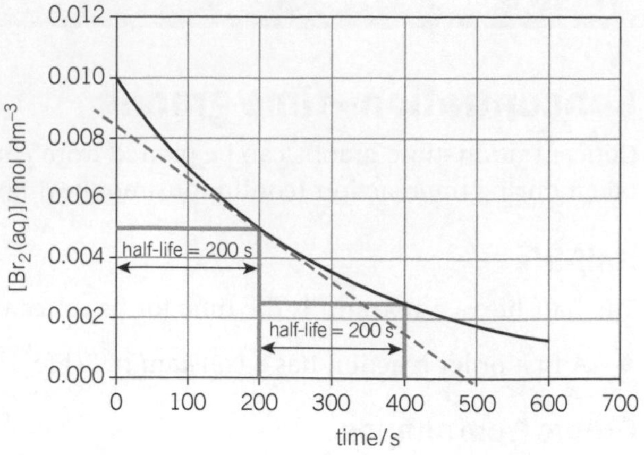

▲ **Figure 3** *Concentration–time graph for the reaction of Br_2 with HCOOH*

Rate constants for first order reactions

For a first order reaction, the rate constant, *k*, can be determined from the constant half-life.

Revision tip

The half-life of a first order reaction is independent of concentration.

Synoptic link

Refer back to Topic 10.1, Reaction rates, to see how the gradient of a tangent gives the rate of reaction.

Synoptic link

The units of a first order reaction are s^{-1}.

See Topic 18.1, Orders, rate equations, and rate constants, for details.

Revision tip

The $k = \ln 2/t_{½}$ method is much easier than the gradient method and overcomes the problems of drawing an accurate tangent. If you compare the value for *k* from the two methods, *k* from half-life is likely to be more accurate.

Method 1 Rate constant from rate.

At any time, the rate can be measured by measuring the gradient of a tangent to the curve. A tangent has been drawn on Figure 3 at $t = 200$ s.

In Figure 3, after 200 s,

- $[Br_2] = 0.0050\ \text{mol dm}^{-3}$
- rate = gradient of tangent $= \frac{0.0082}{490} = 1.673 \times 10^{-5}\ \text{mol dm}^{3}\ \text{s}^{-1}$

rate = $k[Br_2]$

- After 200 s, $k = \frac{\text{rate}}{[Br_2]} = \frac{1.673 \times 10^{-5}}{0.0050} = 3.35 \times 10^{-3}\ \text{s}^{-1}$

Method 2 Rate constant from half-life

For a first order reaction, the rate constant, *k*, can be determined from the constant half-life, $t_{½}$, using the relationship:

$$k = \frac{\ln 2}{t_{½}}$$

For the half-life of 200 s, $k = \frac{0.693}{200} = 3.47 \times 10^{-3}\ \text{s}^{-1}$

Summary questions

1 Sketch the shapes of concentration–time graphs for reactants that are
  a zero order **b** first order. *(2 marks)*

2. The half-life of two first order reactions are: Reaction 1, $t_{1/2} = 21.0$ s; Reaction 2, $t_{1/2} = 4.85 \times 10^{-2}$ s. Using $k = \frac{\ln 2}{t_{½}}$, calculate the rate constant of each reaction. *(2 marks)*

3 The table shows how the concentration of a reactant **A** changes during a reaction.

Time / s	0	360	720	1080	1440
[A] / mol dm^{-3}	0.240	0.156	0.104	0.068	0.045

  a Plot a concentration–time graph. *(4 marks)*
  b By drawing tangents, estimate
    i the initial rate **ii** the rate after 500 s. *(2 marks)*
  c Determine the half-life and show that the reaction is first order with respect to **A**. *(2 marks)*
  d Calculate the rate constant *k*
    i from the rate method in **b** **ii** using $k = \ln 2/t_{½}$. *(2 marks)*

18.3 Rate–concentration graphs and initial rates

Specification reference: 5.1.1

Rate–concentration graphs

Rate–concentration graphs can be plotted from the results of initial rates experiments (see Topic 18.1, Orders, rate equations, and rate constants).

Orders from shapes

Figure 1 shows the shapes of rate–concentration graphs for zero order, first order, and second order.

Determination of rate constants

Figure 2 shows a rate–concentration graph for the first order reactant **A**.

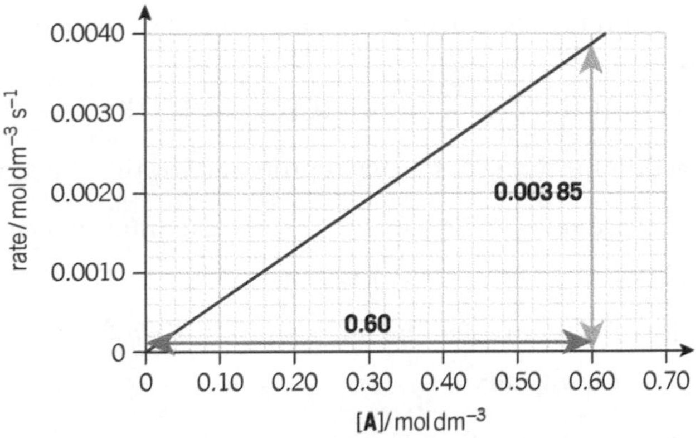

▲ **Figure 2** *Rate–concentration graph of the first order reactant* **A**

From the shape, the reaction is first order with respect to reactant **A**.

rate = $k[\mathbf{A}]$

rate = gradient of straight line = $\dfrac{0.00385}{0.60} = 6.4 \times 10^{-3}\,\mathrm{s}^{-1}$

1 Sketch the shapes of rate–concentration graphs for reactants that are
   a zero order   *(1 mark)*
   b first order.   *(1 mark)*

2 From a rate–concentration graph for a first order reaction, how can you determine the rate constant?   *(1 mark)*

3 The table below shows the results of a clock reaction in terms of a reactant **A**. The reaction is zero order with respect to the other species in the reaction.

[A] /10^{-2} mol dm^{-3}	5.00	4.00	3.00	2.00	1.00
Time, t / s	29	36	52	71	135

   a Calculate the initial rate as $1/t$ for each concentration. *(2 marks)*
   b Plot a rate–concentration graph.   *(4 marks)*
   c Determine
      i the order of reaction with respect to **A**   *(1 mark)*
      ii the rate constant k.   *(2 marks)*

Zero order

rate $\propto [\mathbf{A}]^0$ ∴ rate = constant

- Rate unaffected by changes in concentration

First order

rate $\propto [\mathbf{X}]^1$

- Rate doubles as concentration doubles

Second order

rate $\propto [\mathbf{X}]^2$

- Rate quadruples as concentration doubles

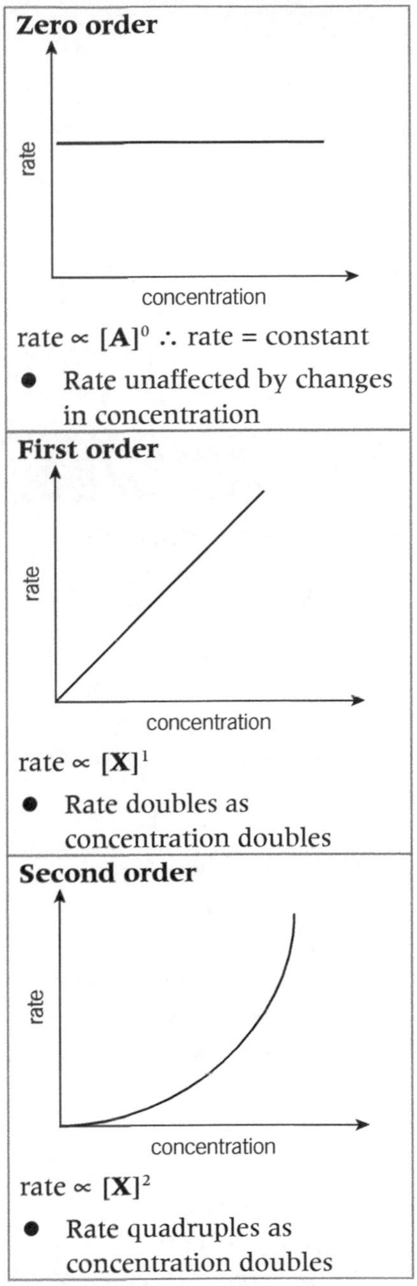

▲ **Figure 1** *Rate–concentration graphs for zero order, first order, and second order*

18.4 Rate-determining step

Specification reference: 5.1.1

Multi-step reactions

Chemical reactions often take place in a series of steps.

- A reaction mechanism summarises the sequence of steps.
- The overall reaction is the sum of the species involved in each step.

Rate-determining step

The **rate-determining step** is the slowest step in the reaction mechanism of a multi-step reaction.

- The rate equation contains the reactants in the rate-determining step.

Predicting a rate equation

A rate equation can be predicted from the rate-determining step.

The overall equation for the reaction of $(CH_3)_3CBr$, with hydroxide ions, OH^- is shown below.

$$(CH_3)_3C–Br + OH^- \rightarrow (CH_3)_3C–OH + Br^-$$

A two-step reaction mechanism has been proposed for this reaction:

Step 1	$(CH_3)_3C–Br \rightarrow (CH_3)_3C^+ + Br^-$	*slow* (rate-determining step)
Step 2	$(CH_3)_3C^+ + OH^- \rightarrow (CH_3)_3C–OH$	*fast*

The rate equation should involve the reactants in the slow step: i.e. **one** $(CH_3)_3C–Br$ molecule.

Therefore, we would predict the rate equation to be:

$$\text{rate} = k[(CH_3)_3C–Br]$$

Predicting possible steps in a reaction mechanism

The steps in a reaction mechanism can be predicted from:

- the rate equation
- the balanced equation for the overall equation.

Hydrogen peroxide, H_2O_2, reacts with iodide ions, I^-, in acid solution.

The **overall** equation is shown below.

$$H_2O_2(aq) + 2H^+(aq) + 2I^-(aq) \rightarrow 2H_2O(l) + I_2(aq)$$

The **rate** equation (determined from experiments) is:

$$\text{rate} = [H_2O_2(aq)][I^-(aq)]$$

The rate equation gives the reactants in the slow rate-determining step:

$H_2O_2(aq) + I^-(aq) \rightarrow$ .... *slow step*

- If this is the first step, there must be further fast steps because the overall equation is different.
- The overall equation results from adding together the species in all the steps.

The following mechanism can be proposed.

First step	$H_2O_2(aq) + I^-(aq) \rightarrow H_2O(l) + IO^-(aq)$	*slow*
Further steps	$IO^-(aq) + H^+(aq) \rightarrow HIO(aq)$	*fast*
	$HIO(aq) + H^+(aq) + I^-(aq) \rightarrow I_2(aq) + H_2O(l)$	*fast*
Overall equation	$H_2O_2(aq) + 2H^+(aq) + 2I^-(aq) \rightarrow 2H_2O(l) + I_2(aq)$	*sum of all steps*

Summary questions

1 a What is meant by the term *rate-determining step*? *(1 mark)*

b What does the rate-determining step tell you about the rate equation? *(1 mark)*

2 A proposed two-step mechanism for a reaction is shown below.

$H_2(g) + ICl(g) \rightarrow HCl(g) + HI(g)$ *slow*

$HI(g) + ICl(g) \rightarrow HCl(g) + I_2(g)$ *fast*

a What is the rate equation for this reaction? *(1 mark)*

b Write the overall equation for this reaction. *(1 mark)*

3 The overall equation for the reaction of carbon monoxide and nitrogen dioxide is:

$CO(g) + NO_2(g) \rightarrow CO_2(g) + NO(g)$

The rate equation is *rate* $= k[NO_2]^2$ and the slow step is the first step. Suggest a possible two-step mechanism for the reaction. *(2 marks)*

18.5 Effect of temperature on rate constants

Specification reference: 5.1.1

The effect of temperature on rate constants

Rate constants can be determined experimentally at different temperatures using the methods described in Topic 18.1, Orders, rate equations, and rate constants, Topic 18.2, Concentration–time graphs, and Topic 18.3, Rate–concentration graphs and initial rates.

Figure 1 shows the increase in rate constant with increasing temperature.

The rate constant is proportional to the rate of reaction. An increase in the rate constant means an increase in the rate.

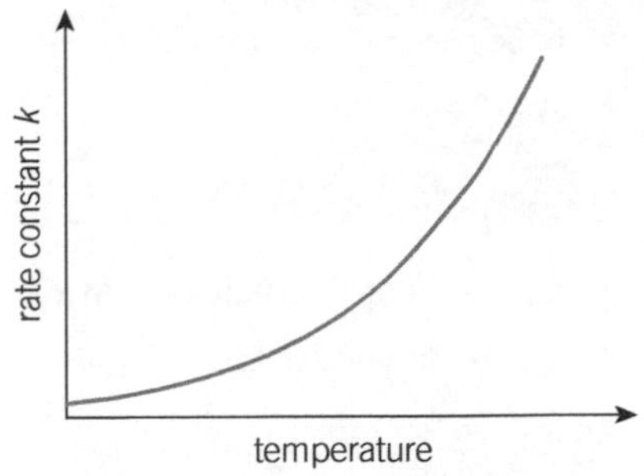

▲ **Figure 1** *Variation of rate constant k with temperature T*

Why does the rate constant increase with temperature?

When temperature is increased:

- the Boltzmann distribution shifts to the right, increasing the proportion of particles that exceed the activation energy, E_a
- the particles move faster and collide more frequently, with more particles colliding with the correct orientation to react.

With increasing temperature, the increase in collision frequency is comparatively small compared with the increased proportion of molecules that exceed E_a from the shift in the Boltzmann distribution.

- A change in rate is mainly determined by E_a.

Synoptic link

Look back at Topic 10.3, The Boltzmann distribution, to revise the basics of the effect of temperature on reaction rates.

The Arrhenius equation

The Arrhenius equation shows how the rate constant is linked to E_a, T, and the pre-exponential factor A (which accounts for collision frequency and correct orientation).

$$k = A\, e^{-E_a/RT}$$

pre-exponential factor (frequency factor) → A

exponential factor (linked to activation energy and temperature) → $e^{-E_a/RT}$

R = gas constant = 8.314 J mol^{-1} K^{-1} T = temperature in Kelvin

Revision tip

The Arrhenius equation and its logarithmic forms are provided on the data sheet.

Determination of E_a and A graphically

The Arrhenius equation can be expressed as a logarithmic relationship:

$$\ln k = -\frac{E_a}{RT} + \ln A$$

A plot of ln k against 1/T gives a straight line graph of the type $y = mx + c$

$$\ln k = -\frac{E_a}{R}\ \frac{1}{T} + \ln A$$

$$y = m\ x + c$$

- gradient $m = -\frac{E_a}{R}$
- intercept on the y axis, $c = \ln A$.

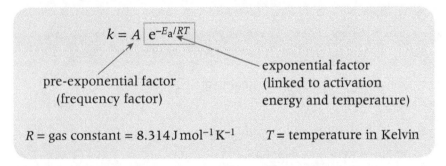

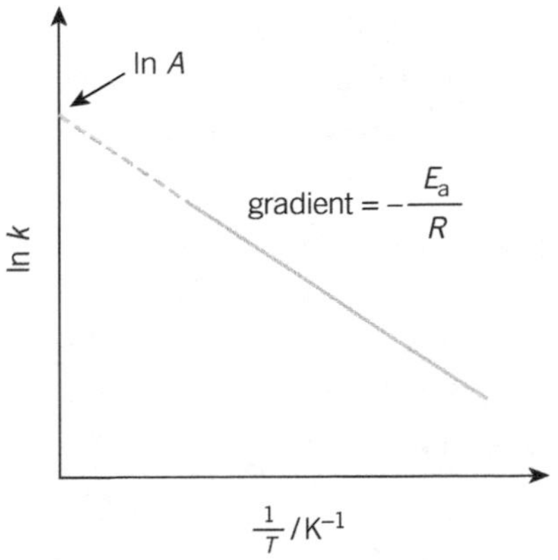

▲ **Figure 2** *Arrhenius plot with ln k positive*

Worked example: Using the Arrhenius equation to determine E_a and A

Table 1 shows rate constants, k, for a reaction at different temperatures.
Plot a suitable graph and determine the activation energy, E_a, and the pre-exponential factor A.

When $\frac{1}{T} = 0$, $\ln k = 25.40$.

▼ **Table 1** *Rate constant k at different temperatures T*

T/K	295	298	305	310	320
k /s^{-1}	0.000 493	0.000 656	0.001 400	0.002 360	0.006 120

Step 1: Calculate the values of ln k and $1/T$ from the data provided.

▼ **Table 2** *ln k and 1/T from values in Table 1*

$\frac{1}{T}$/K^{-1}	0.00 339	0.00 336	0.00 328	0.00 323	0.00 313
ln k	−7.615	−7.329	−6.571	−6.049	−5.096

Step 2: Plot a graph of ln k against $1/T$.

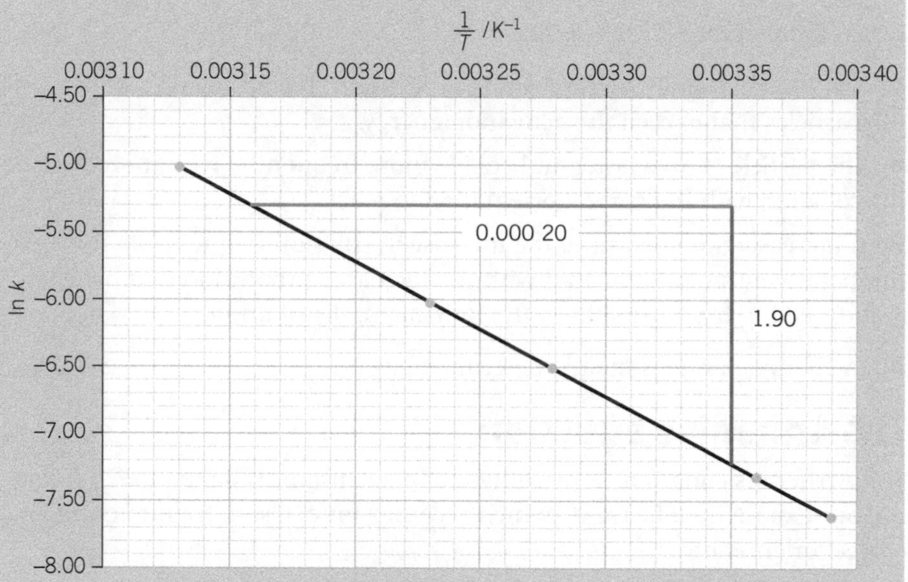

Step 3: Measure the gradient of the straight line and calculate E_a.

$$\text{Gradient} = -\frac{E_a}{R} = \frac{-1.90}{0.00020} = -9500$$

$\therefore E_a = -(8.314 \times -9500)$
$= 78\,983\ \text{J mol}^{-1}$
$= 79\ \text{kJ mol}^{-1}$

Step 4: Calculate A.

The intercept, c, on the y axis at $\frac{1}{T} = 0$ is 25.40

(See information at start of Worked example)

$\ln A = 25.40 \quad \therefore A = 1.07 \times 10^{11}$

Revision tip

On the scale of the graph, the intercept is likely to be well off your graph. In an exam, it is likely that you will be provided with the extrapolation to show the intercept. Alternatively, you will be provided with the value of the intercept.

Revision tip

To work out A from ln A in the Worked example, on your calculator, press SHIFT (OR 2nd), ln, followed by 25.40.

Summary questions

1 What is the effect of temperature on the rate constant and rate of a reaction? *(2 marks)*

2 **a** What graph is plotted to determine the activation energy E_a and pre-exponential factor A? *(1 mark)*

b How are E_a and A determined from the graph? *(2 marks)*

3 The variation of the rate constant with temperature for the first order reaction is given below.

T/K	410	417	426	436
k /s^{-1}	0.0193	0.0398	0.0830	0.2170

a Plot a graph of ln k against $\frac{1}{T}$. *(4 marks)*

b Determine graphically the activation energy of the reaction in kJ mol^{-1}. *(2 marks)*

c The intercept on the y axis is at 36.0. Calculate the value of the pre-exponential factor A. *(1 mark)*

Chapter 18 Practice questions

1 The rate equation for a reaction is: *rate* = k[**A**][**B**]^2.

Which row is consistent with the rate equation?

	[A] /mol dm^{-3}	[B] /mol dm^{-3} s^{-1}	rate /mol dm^{-3} s^{-1}
Given	0.010	0.010	1.2×10^{-4}
A	0.010	0.020	2.4×10^{-4}
B	0.020	0.010	6.0×10^{-4}
C	0.020	0.020	4.8×10^{-4}
D	0.010	0.020	4.8×10^{-4}

(1 mark)

2 A reaction between **A** and **B** has the rate equation: *rate* = k[**A**]^2[**B**].

Which concentration changes could cause the rate to increase by 16 times?

A [**A**] and [**B**] are both doubled.

B [**A**] and [**B**] are increased by 4 times.

C [**A**] is doubled and [**B**] is increased by 4 times.

D [**A**] is increased by 4 times and [**B**] is doubled. *(1 mark)*

3 The half-life of a first order reaction is 12 s.

What is the value, in s^{-1}, of the rate constant?

A 0.058 **C** 12

B 8.3 **D** 17 *(1 mark)*

4 The rate equation is rate = k[**A**][**B**]^2

What are the units of the rate constant k?

A dm^3 mol^{-1} s^{-1} **C** dm^6 mol^{-2} s^{-1}

B mol dm^{-3} s^{-1} **D** mol^2 dm^{-6} s^{-1} *(1 mark)*

5 Which expression gives the value of E_a from the gradient of a ln k against $1/T$ graph?

A –gradient × R

B gradient × R

C –gradient ÷ R

D gradient × R *(1 mark)*

6 The results of a series of experiments are shown below.

Experiment	[A]/mol dm^{-3}	[B]]/mol dm^{-3}	Initial rate/mol dm^{-3} s^{-1}
1	0.0100	0.0100	1.5×10^{-4}
2	0.0200	0.0200	1.20×10^{-3}
3	0.0300	0.0200	1.80×10^{-3}

a Determine the orders. Justify your answer. *(4 marks)*

b What is the rate equation. *(1 mark)*

c Calculate k, including units. *(2 marks)*

d What are the shapes of rate–concentration graphs for **A** and **B**? *(2 marks)*

7 Nitrogen monoxide reacts with ozone by the following mechanism:

Step 1	$NO(g) + O_3(g) \rightarrow NO_2(g) + O_2(g)$	slow
Step 2	$NO_2(g) + O(g) \rightarrow NO(g) + O_2(g)$	fast

a What is the rate equation? *(1 mark)*

b What is the overall equation? *(1 mark)*

c Explain the role of nitrogen monoxide. *(2 marks)*

19.1 The equilibrium constant K_c

Specification reference: 5.1.2

> **Synoptic link**
>
> Look back at Topic 10.5, The equilibrium constant K_c, to revisit the introductory work on equilibrium constants.

> **Revision tip**
>
> Calculating K_c (from Topic 10.5)
>
> $2NO(g) + O_2(g) \rightleftharpoons 2NO_2(g)$,
>
> At equilibrium ($mol\,dm^{-3}$):
>
> $[NO] = 0.100$; $[O_2] = 0.300$;
>
> $[NO_2] = 0.200$.
>
> $$K_c = \frac{[NO_2(g)]^2}{[NO(g)]^2[O_2(g)]}$$
>
> $$= \frac{0.200^2}{0.100^2 \times 0.300} = 13.3\,dm^3\,mol^{-1}$$

> **Synoptic link**
>
> A similar method is used to work out units for rate constants. See Topic 18.1, Orders, rate equations, and rate constants.

> **Revision tip**
>
> In this equilibrium, C(s) is a solid but the other three species are all gases.

> **Revision tip**
>
> Take care when writing K_c expressions. Omit any species that are (s) or (l). This affects both the value of K_c and the units.

> **Revision tip**
>
> Take care with an equilibrium mixture containing miscible liquids. The liquids dissolve in each other, their concentrations can change, and they **are** included in the K_c expression.

The equilibrium constant, K_c

Earlier work on K_c

In Topic 10.5, The equilibrium constant K_c, you learnt how to write an equilibrium constant K_c and to calculate the numerical value for K_c from equilibrium concentrations.

Units of K_c

The units of K_c depend on the number of concentration terms on the top and bottom of the K_c expression. To work out the units:

- substitute concentration units into the expression for K_c
- cancel common units and show the final units on a single line.

e.g. $2NO(g) + O_2(g) \rightleftharpoons 2NO_2(g)$

$$K_c = \frac{[NO_2(g)]^2}{[NO(g)]^2[O_2(g)]}$$

Substitute and cancel units: $\frac{\cancel{(mol\,dm^{-3})^2}}{\cancel{(mol\,dm^{-3})^2}\,(mol\,dm^{-3})}$

Units: $dm^3\,mol^{-1}$

Homogeneous and heterogeneous equilibria

Homogeneous equilibria

A **homogeneous equilibrium** contains equilibrium species that *all* have the same state.

$N_2(g) + 3H_2(g) \rightleftharpoons 2NH_3(g)$ homogeneous: all species are gases

Heterogeneous equilibria

A **heterogeneous equilibrium** contains equilibrium species that have *different* states.

$C(s) + H_2O(g) \rightleftharpoons CO(g) + H_2(g)$ heterogeneous: mixture of states

Writing K_c expressions for heterogeneous equilibria

The concentrations of solids and liquids are essentially constant.

- Species that are (s) or (l) are *omitted* from the K_c expression.
- K_c **only** includes species that are (g) or (aq) – their concentrations *can* change.

In the equilibrium: $C(s) + H_2O(g) \rightleftharpoons CO(g) + H_2(g)$

- $H_2O(g)$, $CO(g)$ and $H_2(g)$ are gases but C(s) is a solid.
- C(s) is constant and is omitted from the K_c expression:

$$K_c = \frac{[CO(g)]\,[H_2(g)]}{[H_2O(g)]}$$ Units = $mol\,dm^{-3}$

Calculation of quantities present at equilibrium

To determine an equilibrium constant experimentally, you need to follow a set method.

- Measure quantities of chemicals for the reaction and mix them together.
- Leave the mixture to allow equilibrium to be set up.
- Measure the equilibrium amount of at least one of the species in the equilibrium mixture.

Worked example: Calculating equilibrium amounts and K_c

Sulfur dioxide and oxygen react to form sulfur trioxide in an equilibrium reaction:

$$2SO_2(g) + O_2(g) \rightleftharpoons 2SO_3(g)$$

0.0250 mol SO_2 and 0.0125 mol O_2 are mixed in a 4.00 dm^3 container. The mixture is allowed to reach equilibrium at constant temperature. The equilibrium mixture contains 0.0050 mol SO_2.

Determine the concentrations of SO_2, O_2, and SO_3 present at equilibrium and calculate K_c.

Step 1: Calculate the amount, in moles, of SO_2 that reacted.

$n(SO_2) = 0.0250 - 0.0050 = \mathbf{0.0200}$ mol

Step 2: Use the equation to find the equilibrium amounts (in mol) of all species in the equilibrium:

equation:	$2SO_2(g)$	+	$O_2(g)$	$\rightleftharpoons$	$2SO_3(g)$
mole ratios	2 mol		1 mol	$\rightarrow$	2 mol
initial moles	0.0250		0.0125		0
change in moles	**−0.0200**		−0.0100		+0.0200
equilibrium amounts	**0.0050**		**0.0025**		**0.0200**

Step 3: Work out the equilibrium concentrations, in mol dm^{-3}.

Volume = 4.00 dm^3. Divide equilibrium moles by 4.00 to find the concentration in mol dm^{-3}.

$[SO_2(g)] = \frac{0.0050}{4.00} = 1.25 \times 10^{-3}\ \text{mol dm}^{-3}$ $\quad [O_2(g)] = \frac{0.0025}{4.00} = 6.25 \times 10^{-4}\ \text{mol dm}^{-3}$

$[SO_3(g)] = \frac{0.0200}{4.00} = 5.00 \times 10^{-3}\ \text{mol dm}^{-3}$

Step 4: Write the expression for K_c, substitute values and calculate K_c.

$$K_c = \frac{[SO_3(g)]^2}{[SO_2(g)]^2[O_2(g)]^2} = \frac{(5.00 \times 10^{-3})^2}{(1.25 \times 10^{-3})^2 \times 6.25 \times 10^{-4}} = 2.56 \times 10^4\ \text{dm}^3\ \text{mol}^{-1}$$

Determination of equilibrium quantities in the laboratory

The method above would be unrealistic in a school or college laboratory but the principle is the same. Equilibria can be set up in solution and the equilibrium amount of one species in the mixture can be found.

Summary questions

1 For each equilibrium, write the expression for K_c, including units.

- **a** $N_2O_4(g) \rightleftharpoons 2NO_2(g)$ *(2 marks)*
- **b** $4HCl(g) + O_2(g) \rightleftharpoons 2H_2O(g) + 2Cl_2(g)$ *(2 marks)*
- **c** $3Fe(s) + 4H_2O(g) \rightleftharpoons Fe_3O_4(s) + 4H_2(g)$ *(2 marks)*

2 N_2 reacts with H_2 in a reversible reaction: $N_2(g) + 3H_2(g) \rightleftharpoons 2NH_3(g)$
At equilibrium, a 200 cm^3 container contains 3.25×10^{-3} mol N_2, 6.25×10^{-2} mol H_2, and 1.64×10^{-3} mol NH_3.

- **a** Calculate the equilibrium concentrations of N_2, H_2, and NH_3. *(3 marks)*
- **b** Calculate K_c to **three** significant figures. Include units. *(3 marks)*

3 CO reacts with H_2 in a reversible reaction: $CO(g) + 2H_2(g) \rightleftharpoons CH_3OH(g)$
0.0250 mol of CO and 0.100 mol H_2 are mixed together in a container of volume 5.00 dm^3.
At equilibrium, 4.50×10^{-3} mol CH_3OH is present.

- **a** Calculate the equilibrium amounts of CO and H_2. *(2 marks)*
- **b** Calculate the equilibrium concentrations of CO(g), $H_2(g)$, and $CH_3OH(g)$. *(3 marks)*
- **c** Calculate K_c to **three** significant figures. Include units. *(3 marks)*

19.2 The equilibrium constant K_p

Specification reference: 5.1.2

Mole fraction and partial pressure

K_p is the equilibrium constant in terms of partial pressures and is commonly used for equilibria involving gases. To calculate K_p, you need to understand mole fractions and partial pressures.

Revision tip

The sum of the mole fractions of the gases in a gas mixture must equal one.

Mole fraction is the same as the proportion by volume of the gas in a gas mixture.

Mole fraction

For a gas **A** in a gas mixture,

$$\text{mole fraction, } x(\mathbf{A}) = \frac{\text{number of moles of } \mathbf{A}}{\text{total number of moles in gas mixture}}$$

Worked example: Calculating mole fractions

A gas mixture contains 4 mol O_2, 10 mol CO_2, and 6 mol H_2.

What is the mole fraction of each gas?

Step 1: Calculate the total number of moles of gas in the mixture.

Total moles = 4 + 10 + 6 = 20 mol

Step 2: Calculate the mole fraction of each gas as $\frac{\text{moles}}{\text{total moles}}$

$x(O_2) = \frac{4}{20} = 0.2$; $x(CO_2) = \frac{10}{20} = 0.5$; $x(H_2) = \frac{6}{20} = 0.3$

Revision tip

Always check that your calculated mole fractions add up to 1.

Here, 0.2 + 0.5 + 0.3 = 1

Partial pressure

The **partial pressure** of a gas, p, is the contribution of the gas to the total pressure, P.

For gas **A** with a mole fraction $x(\mathbf{A})$ in a gas mixture:

- partial pressure, $p(\mathbf{A}) = x(\mathbf{A}) \times P$

Revision tip

The sum of the partial pressures of the gases in a gas mixture **must** equal to the total pressure.

Worked example: Calculating partial pressures

The mole fractions in a gas mixture are $x(O_2) = 0.25$; $x(N_2) = 0.55$; $x(CH_4) = 0.20$.

A gas mixture has a total pressure of 120 kPa.

What are the partial pressures of O_2, N_2, and CH_4?

$p(O_2) = x(O_2) \times P = 0.25 \times 120 = 30$ kPa

$p(N_2) = x(N_2) \times P = 0.55 \times 120 = 66$ kPa

$p(CH_4) = x(CH_4) \times P = 0.20 \times 120 = 24$ kPa

Note that the sum of the partial pressures equals the total pressure: 30 + 66 + 24 = 120 kPa

Revision tip

Always check that your calculated partial pressures add up to the total pressure.

Here, 30 + 66 + 24 = 120

K_p for homogeneous and heterogeneous equilibria

K_p is written in a similar way to K_c, but partial pressures are used instead of concentrations.

K_p only includes gases because only gases have partial pressures. Any other species are ignored.

p is the equilibrium partial pressure in kilopascals (kPa), pascals (Pa), or atmospheres (atm).

Homogeneous equiilbria

For the equilibrium: $H_2(g) + I_2(g) \rightleftharpoons 2HI(g)$

- all species are gases:

$$K_p = \frac{p(HI)^2}{p(H_2) \times p(I_2)}$$

Heterogeneous equiilbria

For the equilibrium: $C(s) + CO_2(g) \rightleftharpoons 2CO(g)$,

- C(s) is solid; $CO_2(g)$ and CO(g) are gases.
- C(s) does not have a partial pressure – only gaseous species are included in the K_p expression:

$$K_p = \frac{p(CO)^2}{p(CO_2)}$$

Worked example: Calculating K_p for a homogeneous equilibrium

$H_2(g)$, CO(g), and $CH_3OH(g)$ form a homogeneous equilibrium (all species are gases):

$$CO(g) + 2H_2(g) \rightleftharpoons CH_3OH(g)$$

An equilibrium mixture contains 12 mol CO(g), 30 mol $H_2(g)$, and 54 mol $CH_3OH(g)$. The total equilibrium pressure is 400 atm. Use this information to calculate K_p.

Step 1: Find the mole fractions of CO, H_2, and CH_3OH

Total number of gas moles = 12 + 30 + 54 = 96 mol

$$x(CO) = \frac{12}{96};\ x(H_2) = \frac{30}{96};\ x(CH_3OH) = \frac{54}{96}$$

Step 2: Find the partial pressures

$$p(CO) = \frac{12}{96} \times 400 = 50\text{ atm};\ p(H_2) = \frac{30}{96} \times 400 = 125\text{ atm};$$

$$p(CH_3OH) = \frac{54}{96} \times 400 = 225\text{ atm}$$

Step 3: Calculate K_p

For the equilibrium: $CO(g) + 2H_2(g) \rightleftharpoons CH_3OH(g)$

$$K_p = \frac{p(CH_3OH)}{p(CO) \times p(H_2)^2} \qquad \text{Units: } K_p = \frac{(atm)}{(atm) \times (atm)^2} = \textbf{atm}^{-2}$$

$$K_p = \frac{225}{50 \times 125^2} = 2.88 \times 10^{-4}\text{ atm}^{-2}$$

Revision tip

Units for K_p:

e.g. p in atm:

$$K_p = \frac{p(HI)^2}{p(H_2) \times p(I_2)} = \frac{atm^2}{atm \times atm} = \text{no units}$$

e.g. p in kPa:

$$K_p = \frac{p(CO)^2}{p(CO_2)}$$

$$\text{units} = \frac{kPa^2}{kPa} = kPa$$

Revision tip

$$x(A) = \frac{\text{number of moles of A}}{\text{total number of moles}}$$

Revision tip

$p(A) = x(A) \times$ total pressure

Check the sum of the partial pressures matches the total pressure:

Partial pressures: 50 + 125 + 225 = 400 atm

Total pressure = 400 atm

Summary questions

1 A gas mixture with a total pressure of 600 kPa contains 15 mol of $Cl_2(g)$, 16 mol of $O_2(g)$, and 9 mol of N_2.

a Calculate the mole fractions. *(3 marks)*

b Calculate the partial pressures. *(3 marks)*

2 For the following equilibria, write an expression for K_p and calculate the value for K_p, including units.

a $2NO_2(g) \rightleftharpoons N_2O_4(g)$
$p(NO_2)$ 150 kPa; $p(N_2O_4)$ 64 kPa *(3 marks)*

b $2CH_4(g) \rightleftharpoons C_2H_2(g) + 3H_2(g)$
$p(CH_4)$ 14.42 atm; $p(C_2H_2)$ 2.52 atm; $p(H_2)$ 1.25 atm *(3 marks)*

c $C(s) + H_2O(g) \rightleftharpoons CO(g) + H_2(g)$
$p(H_2O)$ 600 kPa; $p(CO)$ 30 kPa; $p(H_2)$ 85 kPa *(3 marks)*

3 In the equilibrium: $N_2(g) + 3H_2(g) \rightleftharpoons 2NH_3(g)$:
0.50 mol of $N_2(g)$ is mixed with 1.90 mol $H_2(g)$ and the mixture is allowed to reach equilibrium at constant temperature.
At equilibrium, 0.80 mol of $NH_3(g)$ has formed.

a Calculate the amounts, in mol, of N_2 and H_2 in the equilibrium mixture. *(2 marks)*

b The total equilibrium pressure is 100 atm.
Calculate the mole fractions and partial pressures of the gases in the equilibrium mixture. *(3 marks)*

c Calculate K_p and state its units. *(3 marks)*

19.3 Controlling the position of equilibrium

Specification reference: 5.1.2

The significance of K_c

The numerical value of K_c indicates the position of equilibrium.

- The larger the value of K_c, the further the equilibrium position to the right.
- The smaller the value of K_c, the further the equilibrium position to the left.

Synoptic link

You first encountered the relationship between the value of K and the extent of an equilibrium reaction in Topic 10.5, The equilibrium constant K_c.

The effect of temperature on equilibrium constants

The numerical value of equilibrium constants can be changed **only** by altering the temperature.

The value of an equilibrium constant is **not affected** by changes in concentration, pressure, or the presence of a catalyst.

Revision tip

The same principles apply to K_c and K_p.

Exothermic and endothermic reactions affect the equilibrium constant differently.

Exothermic reactions

In an **exothermic** reaction, the equilibrium constant **decreases** with increasing temperature.

- The equilibrium position shifts to the **left**.
- The equilibrium yield of products **decreases**.

$$H_2(g) + I_2(g) \rightleftharpoons 2HI(g) \quad \Delta H = -9.6\,kJ\,mol^{-1}$$

← Equilibrium position

Temperature/K	K_p
500	160
700	54
1100	25

K_p decreases ↓

Endothermic reactions

In an **endothermic** reaction, the equilibrium constant **increases** with increasing temperature.

- The equilibrium position shifts to the **right**.
- The equilibrium yield of products **increases**.

$$N_2(g) + O_2(g) \rightleftharpoons 2NO(g) \, \Delta H = +180\,kJ\,mol^{-1}$$

→ Equilibrium position

Temperature/K	K_p
500	5×10^{-13}
700	4×10^{-8}
1100	1×10^{-5}

K_p increases ↓

Revision tip

The value of an equilibrium constant is only changed by temperature.

Concentration, pressure, or catalysts have **no effect** on the value of K_c or K_p.

Control of equilibrium position by K_c

At equilibrium $K_c = \frac{[\text{products}]}{[\text{reactants}]}$.

- A change in conditions may move the system out of equilibrium.
- The ratio of $\frac{[\text{products}]}{[\text{reactants}]}$ is then no longer equal to K_c.
- The position of equilibrium then shifts to restore equilibrium.

Changes in temperature

In an **endothermic** reaction, an **increase** in temperature → **increase** in K_c.

- The ratio $\frac{[\text{products}]}{[\text{reactants}]}$ must increase to reach the new larger value of K_c.
- The products **increase** and the reactants **decrease**.
- The position of equilibrium shifts to the **right**.

In an **exothermic** reaction, an **increase** in temperature → **decrease** in K_c.

- The ratio $\frac{[\text{products}]}{[\text{reactants}]}$ must decrease to reach the new smaller value of K_c.
- The products **decrease** and the reactants **increase**.
- The position of equilibrium shifts to the **left**.

Revision tip

Changing the temperature **does** change the value of K_c.

The equilibrium position shifts until the ratio of $\frac{[\text{products}]}{[\text{reactants}]}$ is adjusted to reach the new value of K_c.

Changes in concentration

When the concentration of a **reactant** is increased,

- The ratio $\frac{[\text{products}]}{\mathbf{[reactants]}}$ is now < K_c and the system is no longer in equilibrium:
- Reactants **decrease** and products **increase** to restore the value of K_c.
- The position of equilibrium shifts to the **right**.

If the concentration of a **product** is increased,

- The ratio $\frac{\mathbf{[products]}}{[\text{reactants}]}$ is now > K_c and the system is no longer in equilibrium:
- Products **decrease** and reactants **increase** to restore the value of K_c.
- The position of equilibrium shifts to the **left**.

Changes in pressure

Pressure is proportional to concentration and changes in pressure have a similar effect to changes in concentration.

- If the pressure is increased, the side with more gaseous moles increases more than the side with fewer gaseous moles.
- The ratio $\frac{[\text{products}]}{[\text{reactants}]}$ adjusts to restore the value of K_c.
- The position of equilibrium shifts to the side with fewer gaseous moles.
- If there is no change in the number of gaseous moles, the equilibrium does not need to shift.

Presence of a catalyst

A catalyst has no effect on the value of an equilibrium constant.

- A catalyst only affects reaction rates.
- A catalyst increases the rate of the forward and the reverse reaction by the same amount.
- The position of equilibrium does **not** shift.

Summary questions

1 When temperature is increased, what is the effect on K_c and the equilibrium position for
  a an exothermic reaction *(2 marks)*
  b an endothermic reaction? *(2 marks)*

2 Ammonia, NH_3 is produced in an equilibrium:

$$N_2(g) + 3H_2(g) \rightleftharpoons 2NH_3(g) \quad \Delta H = -93\ \text{kJ mol}^{-1}$$

  a State the effect on K_c of the following changes:
    i increasing temperature *(1 mark)*
    ii removing ammonia as it forms *(1 mark)*
    iii increasing the pressure *(1 mark)*
    iv adding a catalyst. *(1 mark)*
  b Explain the effect on the equilibrium position of the changes in **a** in terms of le Chatelier's principle. *(4 marks)*

3 For the equilibrium in **Q2**, explain the effect on the equilibrium position in terms of K_c on
  a increasing temperature *(3 marks)*
  b removing ammonia as it forms. *(3 marks)*

Revision tip

Changing concentration **does not** change K_c but the equilibrium position still shifts.

The concentration change alters the ratio of $\frac{[\text{products}]}{[\text{reactants}]}$ which then adjusts to restore the value of K_c.

Synoptic link

In Topic 10.4, Dynamic equilibrium and le Chatelier's principle, you used le Chatelier's principle to predict how the equilibrium position shifts during changes in conditions.

Le Chatelier's principle is a useful 'rule of thumb' but the reason why it works is the control of the equilibrium position by equilibrium constants.

Revision tip

There are many different equilibrium constants but the principles learnt for K_c and K_p can be applied to them all.

Chapter 20, Acids, bases, and pH, introduces K_a and K_w, which are variations on K_c with special labels to match their context.

Chapter 19 Practice questions

1 What is the expression for K_c for the equilibrium:
$C(s) + H_2O(g) \rightleftharpoons CO(g) + H_2(g)$?

A $\dfrac{[CO(g)][H_2(g)]}{[C(s)][H_2O(g)]}$ **B** $\dfrac{[C(s)][H_2O(g)]}{[CO(g)][H_2(g)]}$

C $\dfrac{[CO(g)][H_2(g)]}{[H_2O(g)]}$ **D** $\dfrac{[H_2O(g)]}{[CO(g)][H_2(g)]}$

(1 mark)

2 0.300 mol of HBr(g) is added to a 1.00 dm^3 container and left to reach equilibrium at constant temperature: $2HBr(g) \rightleftharpoons H_2(g) + Br_2(g)$.
At equilibrium, 0.0500 mol of both $H_2(g)$ and $Br_2(g)$ are present.
What is the value of K_c?

A 0.0125 **B** 0.0278 **C** 0.0625 **D** 0.250 *(1 mark)*

3 N_2 reacts with H_2 in the equilibrium:

$N_2(g) + 3H_2(g) \rightleftharpoons 2NH_3(g)$ $\Delta H = -93\,kJ\,mol^{-1}$

The temperature is increased. What is the effect on the equilibrium amount of NH_3 and the value of K_c?

	amount of NH_3	value of K_c
A	decrease	increase
B	increase	decrease
C	increase	increase
D	decrease	decrease

(1 mark)

4 SO_2 reacts with O_2 in the equilibrium:

$2SO_2(g) + O_2(g) \rightleftharpoons 2SO_3(g)$ $\Delta H = -197\,kJ\,mol^{-1}$

The pressure is decreased. What is the effect on the equilibrium position and the value of K_c?

	equilibrium position	value of K_c
A	shifts to right	increase
B	shifts to left	decrease
C	shifts to right	no change
D	shifts to left	no change

(1 mark)

5 CO reacts with H_2 to form the equilibrium: $CO(g) + 2H_2(g) \rightleftharpoons CH_3OH(g)$
0.100 mol of CO(g) and 0.200 mol H_2 are mixed in a 500 cm^3 container.
At equilibrium, 80% of CO has reacted.

a Determine the equilibrium concentrations. *(3 marks)*

b Write the expression for K_c. *(1 mark)*

c Calculate K_c, to three significant figures. Include units. *(2 marks)*

d The pressure is doubled.
Explain the effect on the equilibrium position in terms of K_c. *(3 marks)*

6 SO_2 reacts with O_2 to form the equilibrium:

$2SO_2(g) + O_2(g) \rightleftharpoons 2SO_3(g)$ $\Delta H = -198.2\,kJ\,mol^{-1}$

The equilibrium mixture contains 0.15 mol SO_2, 0.45 mol O_2 and 0.90 mol SO_3.
The total pressure is 200 kPa.

a Determine the equilibrium partial pressures. *(3 marks)*

b Write the expression for K_p. *(1 mark)*

c Calculate K_p. Include units. *(2 marks)*

d The temperature is increased.
Explain the effect on the equilibrium position in terms of K_p. *(3 marks)*

20.1 Brønsted–Lowry acids and bases

Specification reference: 5.1.3

Brønsted–Lowry acids and bases

The Brønsted–Lowry model emphasises the role of proton transfer in acid–base reactions.

- A **Brønsted–Lowry acid** is a proton, H^+, donor.
- A **Brønsted–Lowry base** is a proton, H^+, acceptor.

Hydrochloric acid, HCl, is a Brønsted–Lowry acid, releasing H^+, which can be accepted by a base:

$$HCl \rightarrow H^+ + Cl^-$$

A hydroxide ion, OH^-, is a Brønsted–Lowry base, which can accept H^+ from an acid:

$$H^+ + OH^- \rightarrow H_2O$$

Synoptic link

Look back at Topic 4.1, Acids, bases, and neutralisation, to revise ideas about acids, bases, and alkalis, including acids releasing protons and dissociation.

Revision tip

An alkali is a base that dissolves in water forming $OH^-(aq)$ ions. You met the idea of alkalis in Topic 4.1, Acids, bases, and neutralisation. You also met the idea of dissociation.

Remember that 'proton' and H^+ are the same thing and both terms are used.

Conjugate acid–base pairs

An acid dissociates, releasing an H^+ ion and leaving a negative ion.

The dissociation can be shown as an equilibrium:

$$HNO_3(aq) \rightleftharpoons H^+(aq) + NO_3^-(aq)$$

Acid **Base**

- In the forward direction, HNO_3 releases a proton to form its conjugate base NO_3^-.
- In the reverse direction, NO_3^- accepts a proton to forms its conjugate acid HNO_3.
- HNO_3 and NO_3^- are called a **conjugate acid–base pair**.

  A conjugate acid–base pair contains two species that can be interconverted by transfer of a proton.

Revision tip

The formula of the acid and base in a conjugate acid–base pair differs by H^+:

- The conjugate acid donates the H^+.
- The conjugate base is what remains.

Proton transfer in acid–base reactions

An acid–base equilibrium involves two acid–base pairs.

When hydrochloric acid is added to water, proton transfer takes place.

Figure 1 shows the two acid–base pairs in the acid–base equilibrium:

$$HCl(aq) + H_2O(l) \rightleftharpoons H_3O^+(aq) + Cl^-(aq)$$

acid 1 **base 2** **acid 2** **base 1**

▲ **Figure 1** *Acid–base equilibrium of HCl in aqueous solution*

In the forward direction,

- HCl (the conjugate acid) donates a proton to form Cl^- (the conjugate base of HCl).
- HCl and Cl^- are conjugate acid–base pairs.

In the reverse direction,

- H_3O^+ (the conjugate acid) donates a proton to form H_2O (the conjugate base of H_3O^+).
- H_3O^+ and H_2O are conjugate acid–base pairs.

Revision tip

The members of each acid–base pair are on opposite sides of the equilibrium.

When identifying an acid–base pair, start with the acid and look for its base pair on the other side – its formula has H^+ less than the acid.

Clearly label the two acid–base pairs as in Figure 1.

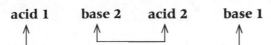

▼ **Table 1** *Mono-, di-, and tribasic acids.*
The replaceable H and the metal in the salt are shown in bold.

Type	Acid	Equation
Monobasic	HNO_3	$HNO_3(aq) + NaOH(aq) \rightarrow$ **Na**$NO_3(aq) + \mathbf{H}_2O(l)$
Dibasic	H_2SO_4	$H_2SO_4(aq) + 2NaOH(aq) \rightarrow$ **Na**$_2SO_4(aq) + 2\mathbf{H}_2O(l)$
Tribasic	H_3PO_4	$H_3PO_4(aq) + 3NaOH(aq) \rightarrow$ **Na**$_3PO_4(aq) + 3\mathbf{H}_2O(l)$

Monobasic, dibasic, and tribasic acids

Monobasic, **dibasic** and **tribasic** refer to the total number of protons in a molecule of the acid that can be replaced in an acid–base reaction.

Table 1 shows some common monobasic, dibasic, and tribasic acids and replacement of protons by a metal ion or ammonium ion to form a salt.

Organic acids do not replace H atoms from the carbon chain.

Synoptic link

See Topic 4.3, Redox, for reactions of metals with acids.

The role of H^+ in acid reactions

In acid reactions, the proton, H^+, is the part of the acid that reacts.

You need to be able to write ionic equations showing the role of H^+ from the acid.

Redox reactions between acids and metals

Dilute acids undergo redox reactions with some metals to form a salt and H_2.

e.g. $2H^+(aq) + Zn(s) \rightarrow Zn^{2+}(aq) + H2(g)$

Reactions between acids and bases

Common bases include carbonates, metal oxides and hydroxides and alkalis. Acids react with bases to form water in a neutralisation reaction.

Neutralisation of acids with carbonates

Carbonates are bases that neutralise acids to form a salt, water, and carbon dioxide.

e.g. $2H^+(aq) + CaCO_3(s) \rightarrow Ca^{2+}(aq) + H_2O(l) + CO_2(g)$

H^+ ions react with aqueous carbonate ions to form CO_2 and H_2O:

$$2H^+(aq) + CO_3^{2-}(aq) \rightarrow CO_2(g) + H_2O(l)$$

Neutralisation of acids with metal oxides or metal hydroxides

An acid is neutralised by a solid metal oxide or hydroxide to form a salt and water only.

e.g. $2H^+(aq) + CuO(s) \rightarrow Cu^{2+}(aq) + H_2O(l)$

Neutralisation of acids with alkalis

With alkalis, the acid and base are in solution.
The ionic equation shows neutralisation of $H^+(aq)$ and $OH^-(aq)$ ions to form water.

$$H^+(aq) + OH^-(aq) \rightarrow H_2O(l)$$

Synoptic link

You first encountered reactions of carbonates, metal oxides, and alkalis with acids in Topic 4.1, Acids, bases, and neutralisation.

Revision tip

For a solid, we still write the ionic equation with the full formula of the carbonate because the metal ions change state during the reaction.

Revision tip

In all these reactions, a solution of a salt is being formed. The anion just depends on the acid:

- Hydrochloric acid forms chlorides.
- Sulfuric acid forms sulfates.
- Nitric acid forms nitrates.

H^+ is the part of an acid that reacts and this role in emphasised in the ionic equation, which is general for any acid.

Summary questions

1 **a** Define the terms:
  i Brønsted–Lowry acid **ii** Brønsted–Lowry base. *(2 marks)*
b State the conjugate base of
  i HI **ii** $HClO_3$ **iii** $CH_3CH_2CH_2COOH$ *(3 marks)*

2 Label the two acid–base pairs in the following acid–base equilibria:
a $CH_3COOH(aq) + OH^-(aq) \rightleftharpoons H_2O(aq) + CH_3COO^-(aq)$ *(1 mark)*
b $HNO_2(aq) + CO_3^{2-}(aq) \rightleftharpoons NO_2^-(aq) + HCO_3^-(aq)$ *(1 mark)*
c $NH_3(aq) + H_2SO_4(aq) \rightleftharpoons NH_4^+(aq) + HSO_4^-(aq)$ *(1 mark)*

3 Write full and ionic equations, with state symbols, for the following reactions:
a aqueous sodium carbonate and hydrochloric acid *(2 marks)*
b solid copper(II) hydroxide and sulfuric acid *(2 marks)*
c magnesium with phosphoric acid, H_3PO_4. *(2 marks)*

20.2 The pH scale and strong acids

Specification reference: 5.1.3

The pH scale

All aqueous solutions contain hydrogen ions, $H^+(aq)$. Different solutions have a very large range of $H^+(aq)$ concentrations with negative powers of 10. pH is a logarithmic scale that compresses this large range of values into numbers that are much easier to understand.

pH as a logarithmic scale

The pH scale in Table 1 shows the relationship between pH and the concentration of $H^+(aq)$.

- A low value of $[H^+(aq)]$ matches a high value of pH.
- A high value of $[H^+(aq)]$ matches a low value of pH.

The mathematical relationships between pH and $[H^+(aq)]$ are:

- $pH = -\log[H^+(aq)]$
- $[H^+(aq)] = 10^{-pH}$

On the logarithmic pH scale, a change of one pH number is equal to a 10 times difference in $[H^+(aq)]$.

- A solution with a pH of 2 has 10 times the $H^+(aq)$ concentration as a solution with a pH of 3.

▼ **Table 1** *pH and hydrogen ion concentrations at 25 °C*

	pH	$[H^+(aq)]$ /mol dm^{-3}
acid	−1	10^1
	0	$10^0 = 1$
	1	10^{-1}
	2	10^{-2}
	3	10^{-3}
	4	10^{-4}
	5	10^{-5}
	6	10^{-6}
neutral	7	10^{-7}
	8	10^{-8}
	9	10^{-9}
	10	10^{-10}
	11	10^{-11}
	12	10^{-12}
	13	10^{-13}
	14	10^{-14}
alkali	15	10^{-15}

Converting between pH and $[H^+(aq)]$

Worked example: Converting from $[H^+(aq)]$ to pH

What is the pH of a solution with a $H^+(aq)$ concentration of 7.56×10^{-5} mol dm^{-3}?

$pH = -\log[H^+(aq)] = -\log(7.56 \times 10^{-5}) = 4.12$

Worked example: Converting from pH to $[H^+(aq)]$

What is the $[H^+(aq)]$ of a solution with a pH of 10.42?

$[H^+(aq)] = 10^{-pH}$ $\therefore$ $[H^+(aq)] = 10^{-10.42} = 3.80 \times 10^{-11}$ mol dm^{-3}

Calculating the pH of a strong monobasic acid

In aqueous solution, a strong monobasic acid, HA, completely dissociates:

e.g. $HCl(aq) \rightarrow H^+(aq) + Cl^-(aq)$

1 mol $\rightarrow$ 1 mol

The concentration of the acid is equal to $[H^+(aq)]$

Worked example: Calculating pH from acid concentration

What is the pH of hydrochloric acid with a concentration of 2.56×10^{-2} mol dm^{-3}?

$[H^+(aq)] = [HCl(aq)] = 2.56 \times 10^{-2}$ mol dm^{-3}

$\therefore pH = -\log[H^+(aq)] = -\log(2.56 \times 10^{-2}) = 1.59$

Revision tip

On calculators, the log button is shown as 'log' or 'lg'.

It is good practice to give your pH answers to two decimal places.

Check your answer. Is it sensible? The pH should be within one of the negative power of $[H^+(aq)]$.

Revision tip

Calculators have a 'SHIFT' or '2nd' button to access the '10^x' function above the 'log' key. The SHIFT key is usually positioned at the top left of the keyboard.

Synoptic link

For details of dissociation and the strength of acids, see Topic 4.1, Acids, bases, and neutralisation.

Revision tip

For a strong monobasic acid, $[H^+(aq)]$ is the same as the acid concentration.

The pH of a strong monobasic acid can be calculated directly from the concentration of the acid.

Worked example: Calculating acid concentration from pH

What is the concentration of hydrochloric acid with a pH of 2.32?

$[H^+(aq)] = 10^{-pH} = 10^{-2.32} = 4.79 \times 10^{-3}\ mol\ dm^{-3}$

HCl is a strong monobasic acid and completely dissociates.

$\therefore [HCl(aq)] = [H^+(aq)] = 4.79 \times 10^{-3}\ mol\ dm^{-3}$

Summary questions

1 **a** Calculate the pH for the following concentrations of H^+ ions.

- **i** $1 \times 10^{-3}\ mol\ dm^{-3}$ *(1 mark)*
- **ii** $1 \times 10^{-12}\ mol\ dm^{-3}$ *(1 mark)*
- **iii** $0.000\,000\,1\ mol\ dm^{-3}$ *(1 mark)*

b Calculate $[H^+(aq)]$ for solutions with the following pH values.

- **i** pH 5 *(1 mark)*
- **ii** pH 8 *(1 mark)*

c **i** How many times more hydrogen ions are in a 1 dm^3 solution of pH 3 than a 1 dm^3 solution of pH 9? *(1 mark)*

ii A solution has 100 000 000 times fewer hydrogen ions than a solution of pH 5. Both solutions have the same volume. What is the pH of the first solution? *(1 mark)*

2 **a** Calculate the pH, to two decimal places, of solutions with the following $H^+(aq)$ concentrations.

- **i** $2.50 \times 10^{-3}\ mol\ dm^{-3}$ *(1 mark)*
- **ii** $8.10 \times 10^{-6}\ mol\ dm^{-3}$ *(1 mark)*
- **iii** $3.72 \times 10^{-8}\ mol\ dm^{-3}$ *(1 mark)*

b Calculate the $H^+(aq)$ concentrations in solutions with the following pH values.

- **i** pH 3.81 *(1 mark)*
- **ii** pH 8.72 *(1 mark)*
- **ii** pH 10.76 *(1 mark)*

3 **a** 25 cm^3 of 0.0200 $mol\ dm^{-3}$ hydrochloric acid is diluted with water and the solution made up to 100 cm^3.
What is the pH of the diluted acid? *(2 marks)*

b 0.73 g of hydrogen chloride gas is dissolved in water to prepare 250 cm^3 of hydrochloric acid.
What is the pH of the hydrochloric acid? *(3 marks)*

20.3 The acid dissociation constant K_a

Specification reference: 5.1.3

Strong and weak acids

In aqueous solution:

- strong acids **completely** dissociate,
  e.g. $HCl(aq) \rightarrow H^+(aq) + Cl^-(aq)$
- weak acids **partially** dissociate,
  e.g. $CH_3COOH(aq) \rightleftharpoons H^+(aq) + CH_3COO^-(aq)$

Synoptic link

For more details of strong and weak acids, see Topic 4.1, Acids, bases, and neutralisation.

The acid dissociation constant K_a

The acid dissociation constant, K_a, is the equilibrium constant that measures the extent of acid dissociation.

For a weak acid HA:

$$HA(aq) \rightleftharpoons H^+(aq) + A^-(aq)$$

$$K_a = \frac{[H^+(aq)]\,[A^-(aq)]}{[HA(aq)]} \quad \text{Units: mol dm}^{-3}$$

- The larger the value of K_a, the greater the dissociation and the stronger the acid.

The K_a expression of any weak acid can be written in a similar way:

e.g. $CH_3COOH(aq) \rightleftharpoons H^+(aq) + CH_3COO^-(aq)$

$$K_a = \frac{[H^+(aq)]\,[CH_3COO^-(aq)]}{[CH_3COOH(aq)]}$$

Synoptic link

K_a is essentially the same as K_c but is used specifically for the dissociation of acids. For more details about K_c, see Chapter 19, Equilibrium.

K_a and pK_a

Different weak acids have different K_a values, with negative indices.

As with pH, pK_a is a logarithmic scale which converts K_a values into more manageable numbers.

The mathematical relationship between pK_a and K_a is:

- $pK_a = -\log K_a$
- $K_a = 10^{-pKa}$

Synoptic link

This logarithmic relationship is the same idea as with pH and $[H^+(aq)]$ that you met in Topic 20.2, The pH scale and strong acids.

Worked example: Converting from K_a to pK_a

What is the pK_a value of a weak acid with a K_a value of 7.24×10^{-3} mol dm^{-3}?

$pK_a = -\log K_a \qquad \therefore pK_a = -\log(7.24 \times 10^{-3}) = 2.14$ (to two decimal places)

Revision tip

Check your answer. Is it sensible? pK_a should be within one of the negative power of K_a.

In this example, the negative power of K_a is −3 and the calculate pK_a value is 2.14.

Worked example: Converting from pK_a to K_a

What is the K_a value of a weak acid with a pK_a of 6.24?

$K_a = 10^{-pKa} \qquad \therefore [H^+(aq)] = 10^{-6.24} = 5.75 \times 10^{-7}$

Revision tip

'Strong' and 'weak' are terms that describe the extent of dissociation.

'Concentrated' and 'dilute' describe the number of moles dissolved in a volume of solution.

Revision tip

As with all equilibrium constants, the value of K_a changes with temperature (K_a increases with temperature).

The values given here are correct at 25 °C.

Comparing K_a and pK_a values

- The stronger the acid, the larger the K_a value and the smaller the pK_a value.
- The weaker the acid, the smaller the K_a value and the larger the pK_a value.

Table 1 compares the K_a and pK_a values of three acids.

▼ **Table 1** *K_a and pK_a values and relative acid strength*

Acid		K_a / mol dm^{-3}	pK_a	Relative acid strength
Nitrous acid	HNO_2	4.10×10^{-4}	3.39	strongest acid
Methanoic acid	HCOOH	1.77×10^{-4}	3.75	
Ethanoic acid	CH_3COOH	1.76×10^{-5}	4.75	weakest acid

Summary questions

1 For the acid–base equilibria below, write expressions for K_a.

a $CH_3CH_2COOH(aq) \rightleftharpoons H^+(aq) + CH_3CH_2COO^-(aq)$ *(1 mark)*

b $HNO_2(aq) \rightleftharpoons H^+(aq) + NO_2^-(aq)$ *(1 mark)*

c $H_2SO_3(aq) \rightleftharpoons H^+(aq) + HSO_3^-(aq)$ *(1 mark)*

2 **a** Calculate pK_a from the following K_a values.
Give your answers to two decimal places.

i 1.38×10^{-3} mol dm^{-3} *(1 mark)*

ii 3.16×10^{-7} mol dm^{-3} *(1 mark)*

iii 5.75×10^{-10} mol dm^{-3} *(1 mark)*

b Calculate K_a for the following pK_a values.
Give answers to two decimal places.

i pK_a 2.12 *(1 mark)*

ii pK_a 3.17 *(1 mark)*

ii pK_a 4.47 *(1 mark)*

3 **a** The pK_a values of HBrO and HCN are 8.70 and 9.21 respectively.
Write equations for the dissociation of each acid, and explain which is the stronger acid. *(2 marks)*

b $ClCH_2COOH$ is a stronger acid than C_6H_5COOH. When the two acids are mixed together, an acid–base equilibrium is set up and the stronger acid donates a proton to the weaker acid.
Write an equation for the acid–base equilibrium and identify the two acid–base pairs. *(1 mark)*

20.4 The pH of weak acids

Specification reference: 5.1.3

Calculations of pH and K_a for weak acids

Dissociation of a weak acid and pH

For a weak monobasic acid, HA, in aqueous solution:

$$HA(aq) \rightleftharpoons H^+(aq) + A^-(aq) \qquad K_a = \frac{[H^+(aq)]\,[A^-(aq)]}{[HA(aq)]}$$

$[H^+(aq)]$ and pH depend upon:

- the acid concentration, $[HA(aq)]$
- the acid dissociation constant, K_a.

Equilibrium concentrations and approximations

The equilibrium concentration of $H^+(aq)$

- The dissociation of HA produces $H^+(aq)$ and $A^-(aq)$ ions in equal amounts.
- The dissociation of water is negligible and we can ignore any additional $H^+(aq)$ ions from water.

$\therefore$ at equilibrium, $[H^+(aq)]_{equilibrium} \sim [A^-(aq)]_{equilibrium}$

The equilibrium concentration of HA(aq)

- Only a very small proportion of HA molecules dissociate into H^+ and A^- ions.
- Any decrease in $[HA(aq)]$ during dissociation can be neglected.

$\therefore [HA]_{equilibrium} \sim [HA]_{undissociated}$

Combining the two approximations,

$$K_a = \frac{[H^+(aq)]\,[A^-(aq)]}{[HA(aq)]} = \frac{[H^+(aq)]^2}{[HA(aq)]}$$

Rearranging the equation:

$$[H^+(aq)] = \sqrt{(K_a \times [HA(aq)])}$$

Worked example: Calculating the pH of a weak acid

Calculate the pH of 0.0125 mol dm^{-3} methanoic acid, HCOOH, at 25 °C.
$K_a = 1.70 \times 10^{-4}$ mol dm^{-3}.

Step 1: Calculate $[H^+(aq)]$ from K_a and $[HA(aq)]$

HCOOH is a weak acid and partially dissociates: $HCOOH(aq) \rightleftharpoons H^+(aq) + HCOO^-(aq)$

$$K_a = \frac{[H^+(aq)]\,[HCOO^-(aq)]}{[HCOOH(aq)]} \sim \frac{[H^+(aq)]^2}{[HCOOH(aq)]}$$

$$[H^+(aq)]^2 = K_a \times [HCOOH(aq)]$$

$$\therefore [H^+(aq)] = \sqrt{(K_a \times [HCOOH(aq)])} = \sqrt{(1.70 \times 10^{-4}) \times 0.0125}$$

$$= 1.46 \times 10^{-3} \text{ mol dm}^{-3}$$

Step 2: Calculate pH from $[H^+(aq)]$

$$pH = -\log[H^+(aq)] = -\log(1.46 \times 10^{-3}) = 2.84$$

Worked example: Calculating K_a for a weak acid

The pH of a 0.0109 mol dm^{-3} solution of a weak acid, HA, is 3.39. Calculate K_a.

Step 1: Use your calculator to find $[H^+(aq)]$

$$[H^+(aq)] = 10^{-pH} = 10^{-3.39} = 4.07 \times 10^{-4} \text{ mol dm}^{-3}$$

Step 2: Calculate K_a from $[H^+(aq)]$ and $[HA(aq)]$

$$K_a = \frac{[H^+(aq)]\,[A^-(aq)]}{[HA(aq)]} \sim \frac{[H^+(aq)]^2}{[HA(aq)]} = \frac{(4.07 \times 10^{-4})^2}{0.0109} = 1.52 \times 10^{-5} \text{ mol dm}^{-3}$$

Revision tip

Compare this with a strong acid which totally dissociates:

$[H^+(aq)]$ and pH depend **only** on the acid concentration, $[H^+(aq)]$.

For details, see Topic 20.2, The pH scale and strong acids.

Synoptic link

See Topic 20.3, The acid dissociation constant K_a, for details of the dissociation of weak acids.

Synoptic link

See Topic 20.5, pH and strong bases, for details of the dissociation of water.

Revision tip

The key to calculating the pH of a weak acid is the relationship:

$[H^+(aq)] = \sqrt{(K_a \times [HA(aq)])}$

It is well worth memorising this equation!

If you know two of $[H^+(aq)]$ (or pH), K_a, and $[HA(aq)]$, you can calculate the third quantity.

Synoptic link

It will be useful to refer back to the start of this topic where the two approximations are introduced.

Limitations of pH and K_a calculations for weak acids

The two approximations introduced at the start of this topic greatly simplify pH calculations of weak acids but they break down under certain situations and can introduce significant calculation errors.

Approximation: $[H^+]_{equilibrium} \sim [A^-]_{equilibrium}$

This approximation breaks down for very weak acids or very dilute solutions.

Approximation: $[HA]_{equilibrium} \sim [HA]_{undissociated}$

This approximation breaks down for 'stronger' weak acids and for concentrated solutions.

Summary questions

1 In pH calculations of weak acids, two approximations are often used.
  - **a** Explain the approximation for the concentration of $H^+(aq)$. *(1 mark)*
  - **b** Explain the approximation for the concentration of HA. *(1 mark)*

2 **a** Find the pH of solutions of a weak acid HA ($K_a = 6.46 \times 10^{-5}$ mol dm^{-3}), with the following concentrations.
  - **i** 1.00 mol dm^{-3} *(2 marks)*
  - **ii** 0.250 mol dm^{-3} *(2 marks)*
  - **iii** 4.20×10^{-3} mol dm^{-3} *(2 marks)*

  b A 0.125 mol dm^{-3} solution of a weak acid HA has a pH of 2.32. Calculate K_a. *(2 marks)*

  c A solution of HNO_2 has a pH of 3.64. Calculate $[HNO_2(aq)]$. ($K_a = 4.00 \times 10^{-4}$ mol dm^{-3}). *(2 marks)*

3 **a** Under which situations do the two approximations used in weak acid pH calculations break down? *(2 marks)*

  b A 0.0250 mol dm^{-3} solution of the weak acid HCN has a pK_a of 9.31. Calculate the concentration of $CN^-(aq)$ ions in the solution. *(3 marks)*

20.5 pH and strong bases

Specification reference: 5.1.3

The ionic product of water K_w

Water dissociates very slightly, acting as a weak acid:

$$H_2O(l) \rightleftharpoons H^+(aq) + OH^-(aq)$$

K_w is called the **ionic product of water**:

$$K_w = [H^+(aq)]\,[OH^-(aq)]$$

The value of K_w at 25 °C is $1.00 \times 10^{-14}\,mol^2\,dm^{-6}$

The pH of water and aqueous solutions

In all aqueous solutions, $H^+(aq)$ and $OH^-(aq)$ ions are both present. The numerical concentrations of $H^+(aq)$ and OH^- are controlled by the value of K_w.

- When $[H^+(aq)] = [OH^-(aq)]$ a solution is neutral.
- When $[H^+(aq)] > [OH^-(aq)]$ a solution is acidic.
- When $[H^+(aq)] < [OH^-(aq)]$ a solution is alkaline.

The pH of strong bases

Potassium hydroxide, KOH, is a strong monobasic base and completely dissociates in water:

$$KOH(aq) \rightarrow K^+(aq) + OH^-(aq)$$

Therefore the $[KOH(aq)]$ is equal to $[OH^-(aq)]$

Worked example: Calculating the pH of a solution of a strong base

A solution of KOH, has a concentration of $5.75 \times 10^{-2}\,mol\,dm^{-3}$. What is the pH at 25 °C?

Step 1: Use K_w and $[OH^-(aq)]$ to find $[H^+(aq)]$

$$K_w = [H^+(aq)]\,[OH^-(aq)] = 1.00 \times 10^{-14}\,mol^2\,dm^{-6}$$

$$\therefore [H^+(aq)] = \frac{K_w}{[OH^-(aq)]} = \frac{1.00 \times 10^{-14}}{5.75 \times 10^{-2}} = 1.74 \times 10^{-13}\,mol\,dm^{-3}$$

Step 2: Use your calculator to find pH

$$pH = -\log[H^+(aq)] = -\log(1.74 \times 10^{-13}) = 12.76$$

Synoptic link

The dissociation of water is endothermic and K_w increases with increasing temperature. For the reason, see Topic 19.3, Controlling the position of equilibrium.

Revision tip

$[H^+(aq)]\,[OH^-(aq)]$ is always equal to $K_w = 1.00 \times 10^{-14}\,mol^2\,dm^{-6}$ at 25 °C.

Revision tip

In a neutral solution at 25 °C,

$[H^+(aq)]$ and $[OH^-(aq)]$ both have a concentration of $1.00 \times 10^{-7}\,mol\,dm^{-3}$

- $pH = -\log(1 \times 10^{-7}) = 7$

Revision tip

The pH of a strong base can be calculated from:

- the concentration of the base
- the ionic product of water, K_w.

Revision tip

For a strong monobasic base, $[OH^-(aq)]$ is the same as the base concentration.
Here, $[OH^-(aq)] = [KOH(aq)]$

Summary questions

1 a Write the expression for the ionic product of water. *(1 mark)*

b Calculate $[H^+(aq)]$ and $[OH^-(aq)]$ in solutions with the following pH values.

i pH = 4.00 *(2 marks)* ii pH = 8.25 *(2 marks)*

2 a Calculate the pH of the following solutions at 25 °C:

i $0.0191\,mol\,dm^{-3}$ NaOH(aq) *(2 marks)*

ii $4.19 \times 10^{-3}\,mol\,dm^{-3}$ KOH (aq) *(2 marks)*

b Calculate the concentration of the following strong bases at 25 °C:

i NaOH(aq) with a pH of 11.23 *(2 marks)*

ii KOH(aq) with a pH of 13.64 *(2 marks)*

3 a Calculate the pH of a solution of $2.49 \times 10^{-3}\,mol\,dm^{-3}$ $Ca(OH)_2$ at 25 °C. *(3 marks)*

b Calculate the pH of the following at 50 °C. (K_w at 50 °C = $5.48 \times 10^{-14}\,mol^2\,dm^{-6}$.)

i water. *(2 marks)* ii $0.0125\,mol\,dm^{-3}$ NaOH. *(2 marks)*

Chapter 20 Practice questions

1 An acid–base equilibrium is shown below.

$HSO_4^-(aq) + H_2O(l) \rightleftharpoons H_3O^+(aq) + SO_4^{2-}(aq)$

Which statement is correct?

A H_2O is the conjugate base of HSO_4^-

B H_3O^+ is the conjugate acid of H_2O

C SO_4^{2-} is the conjugate base of H_3O^+

D HSO_4^- is the conjugate base of SO_4^{2-} *(1 mark)*

2 A solution of sodium hydroxide has pH of 12.0. The solution is made 10 times more dilute.

What is the pH of the solution?

A 11.0 **B** 11.3

C 12.7 **D** 13.0 *(1 mark)*

3 A 0.080 mol dm^{-3} solution of a weak acid HA has a pH of 4.51.

What is the value of K_a in mol dm^{-3}?

A 9.55×10^{-10} **B** 1.19×10^{-8}

C 3.09×10^{-5} **D** 3.86×10^{-4} *(1 mark)*

4 Calcium hydroxide, $Ca(OH)_2$ completely dissociates in water.

What is the pH of 1.00×10^{-4} mol dm^{-3} $Ca(OH)_2$ at 25 °C?

A 3.7 **B** 4.0

C 10.0 **D** 10.3 *(1 mark)*

5 **a** Calculate the pH of 0.125 mol dm^{-3} solutions of the following.

i HCl *(1 mark)*

ii NaOH *(2 marks)*

b 50.0 cm^3 of 0.125 mol dm^{-3} HCl is diluted with an equal volume of water.

What is the pH of the diluted 100 cm^3 solution? *(2 marks)*

c 20.0 cm^3 of 0.0800 mol dm^{-3} NaOH is mixed with 80.0 cm^3 of 0.0600 mol dm^{-3} HCl.

What is the pH of the resulting solution? *(5 marks)*

6 Propanoic acid, CH_3CH_2COOH is a weak Brønsted–Lowry acid ($pK_a = 4.87$).

a Define the following terms

i Brønsted–Lowry acid *(1 mark)*

ii weak acid. *(1 mark)*

b When propanoic acid is added to water, an acid–base equilibrium is set up.

Write an equation for this equilibrium and label the acid–base pairs. *(2 marks)*

c Write full and ionic equations, with state symbols, for the reaction of propanoic acid with solid calcium carbonate. *(2 marks)*

d **i** Write the K_a expression for propanoic acid. *(1 mark)*

ii Calculate the pH of a 0.125 mol dm^{-3} solution of propanoic acid. *(3 marks)*

iii Calculations of the pH of weak acids make two approximations.

Under what situations do the two pH approximations break down? *(2 marks)*

21.1 Buffer solutions

Specification reference: 5.1.3

Buffer solutions

Buffer solutions are used to minimise changes in pH to a solution on addition of small amounts of an acid or base.

Acid buffer solutions contain two components, a weak acid HA, and its conjugate base A^-:

$$HA(aq) \rightleftharpoons H^+(aq) + A^-(aq)$$

weak acid — conjugate base

- The weak acid HA removes added alkali.
- The conjugate base A^- removes added acid.

> **Revision tip**
> As the buffer works, the pH does change but only by a small amount – you should not assume that the pH stays completely constant.

> **Synoptic link**
> You first learnt about conjugate acid–base pairs in Topic 20.1, Brønsted–Lowry acids and bases.

Preparing buffer solutions

You need to know two methods for preparing buffer solutions.

Each method produces a buffer solution that contains

- a large concentration of the weak acid HA
- a large concentration of the conjugate base A^-

> **Revision tip**
> In a buffer solution, HA and A^- act as two reservoirs to remove added acid and alkali:
> - HA removes alkalis
> - A^- removes acids.

Preparing a buffer from a weak acid and its salt

A buffer solution can be prepared by mixing together

- a solution of a weak acid: e.g. ethanoic acid, CH_3COOH
- a solution of a salt of the weak acid: e.g. sodium ethanoate, CH_3COONa.

Preparing a buffer by partial neutralisation of a weak acid

A buffer solution can also be prepared by mixing together

- an excess of a solution of a weak acid: e.g. $CH_3COOH(aq)$
- a solution of an alkali: e.g. $NaOH(aq)$.

The weak acid is partially neutralised by the alkali, forming its conjugate base, and leaving some of the weak acid unreacted.

> **Synoptic link**
> For details of equilibrium position and le Chatelier's principle, see Topic 10.4, Dynamic equilibrium and le Chatelier's principle.

The role of the acid–base pair in a buffer solution

The equilibrium in an acid buffer solution is shown below.

$$HA(aq) \rightleftharpoons H^+(aq) + A^-(aq)$$

- The control of pH by a buffer solution can be explained in terms of shifts in the equilibrium position, using le Chatelier's principle.

The conjugate base removes added acid

On addition of an acid, $H^+(aq)$:

- $H^+(aq)$ ions react with the conjugate base $A^-(aq)$.
- The equilibrium position shifts to the left, removing most of the $H^+(aq)$ ions.

> **Revision tip**
> Don't be fooled into thinking that NaOH is the conjugate base in the buffer. NaOH is being used to **make** the conjugate base A^- by reacting with some of the weak acid HA.

The weak acid removes added alkali

On addition of an alkali, $OH^-(aq)$:

- The small concentration of $H^+(aq)$ in the equilibrium reacts with $OH^-(aq)$ ions:

  $H^+(aq) + OH^-(aq) \rightarrow H_2O(l)$

added alkali

$$HA(aq) \rightleftharpoons H^+(aq) + A^-(aq)$$

added acid

▲ **Figure 1** *Shifting the buffer equilibrium. Additions of acid and alkali shift the equilibrium position in opposite directions*

$$[H^+(aq)] = K_a \times \frac{[HA(aq)]}{[A^-(aq)]}$$

acid dissociation constant

ratio of weak acid and its conjugate base

▲ **Figure 2** *The rearranged K_a expression used for calculating $[H^+(aq)]$ and then pH*

Revision tip

Take care. Weak acid and buffer pH calculations use the same K_a expression but the methods are different.

For weak acids, $[H^+] = [A^-]$

For buffers, $[H^+] \neq [A^-]$

Revision tip

If you are given pK_a and the concentrations of HA and A^- are the same, pH is equal to the pK_a value.

- HA(aq) dissociates to restore most of H^+(aq) ions, shifting the equilibrium position to the right.

Figure 1 summarises the shifts in equilibrium position on addition of an acid and an alkali.

Calculating the pH of buffer solutions

The pH of a buffer solution depends upon:

- the K_a value of HA(aq)

- the concentration ratio, $\frac{[HA(aq)]}{[A^-(aq)]}$

The equilibrium and K_a expression for a weak acid buffer solution are shown below:

$$HA(aq) \rightleftharpoons H^+(aq) + A^-(aq) \qquad K_a = \frac{[H^+(aq)]\,[A^-(aq)]}{[HA(aq)]}$$

To work out the pH of a buffer solution, you rearrange the K_a expression as shown in Figure 2.

🖩 Worked example: Calculating the pH of a buffer solution

A buffer solution has concentrations:

$0.500\,mol\,dm^{-3}$ HCOOH and $0.250\,mol\,dm^{-3}$ HCOONa.

Calculate the pH. $\qquad K_a(HCOOH) = 1.78 \times 10^{-4}\,mol\,dm^{-3}$

Step 1: Calculate $[H^+(aq)]$ from K_a, $[HA(aq)]$, and $[A^-(aq)]$.

$$[H^+(aq)] = K_a \times \frac{[HCOOH(aq)]}{[HCOO^-(aq)]} = 1.78 \times 10^{-4} \times \frac{0.500}{0.250}$$

$$= 3.56 \times 10^{-4}\,mol\,dm^{-3}$$

Step 2: Calculate pH from $[H^+(aq)]$.

$$pH = -\log[H^+(aq)] = -\log(3.56 \times 10^{-4}) = 3.45$$

Summary questions

1. A buffer solution is prepared based on methanoic acid, HCOOH.
   a. State two methods for preparing this buffer solution. *(2 marks)*
   b. Explain in terms of equilibrium how the buffer solution removes added acid and added alkali. *(4 marks)*

2. Calculate the pH of the buffer solutions with the concentrations below.
   a. $0.200\,mol\,dm^{-3}$ C_2H_5COOH and $0.800\,mol\,dm^{-3}$ C_2H_5COONa $(K_a = 1.34 \times 10^{-5}\,mol\,dm^{-3})$. *(2 marks)*
   b. $1.25\,mol\,dm^{-3}$ C_6H_5COOH and $0.250\,mol\,dm^{-3}$ C_6H_5COONa $(K_a = 6.28 \times 10^{-5}\,mol\,dm^{-3})$. *(2 marks)*

3. K_a of $CH_3COOH = 1.74 \times 10^{-5}\,mol\,dm^{-3}$ at 25 °C
   a. A buffer solution is prepared by mixing $600\,cm^3$ of $0.720\,mol\,dm^{-3}$ CH_3COOH with $400\,cm^3$ $0.300\,mol\,dm^{-3}$ CH_3COONa. Calculate the pH of the buffer solution. *(4 marks)*
   b. A buffer solution is prepared by adding $100\,cm^3$ $0.500\,mol\,dm^{-3}$ NaOH(aq) to $400\,cm^3$ $0.200\,mol\,dm^{-3}$ CH_3COOH. Calculate the pH of the buffer solution. *(6 marks)*

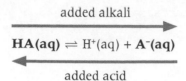

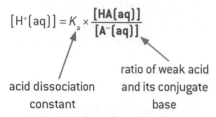

21.2 Buffer solutions in the body

Specification reference: 5.1.3

The carbonic acid–hydrogencarbonate buffer system

A healthy human body maintains pH values within narrow ranges in different parts of the body. The pH in the body is controlled by buffer solutions.

Blood plasma is maintained within the pH range 7.35–7.45, mainly by the carbonic acid–hydrogencarbonate (H_2CO_3/HCO_3^-) buffer system.

- Carbonic acid, $H_2CO_3(aq)$, is the weak acid.
- Hydrogencarbonate, $HCO_3^-(aq)$, is the conjugate base of $H_2CO_3(aq)$.

In the control of pH,

- $H_2CO_3(aq)$ removes excess alkali from the blood.
- $HCO_3^-(aq)$ removes excess acid from the blood.

The equilibrium in the carbonic acid–hydrogencarbonate (H_2CO_3/ HCO_3^-) buffer system is shown below.

$$\underset{\text{weak acid}}{H_2CO_3(aq)} \rightleftharpoons \underset{\text{conjugate base}}{H^+(aq) + HCO_3^-(aq)}$$

- The control of pH by this buffer system can be explained in terms of shifts in the equilibrium position using le Chatelier's principle.
- Figure 1 summarises the shifts in equilibrium position on addition of an acid and an alkali.

Synoptic link

For details of equilibrium position and le Chatelier's principle, see Topic 10.4, Dynamic equilibrium and le Chatelier's principle. The principle here is the same as covered in Topic 21.1, Buffer solutions.

added alkali →

$$\mathbf{H_2CO_3(aq)} \rightleftharpoons H^+(aq) + \mathbf{HCO_3^-(aq)}$$

← added acid

▲ **Figure 1** *The carbonic acid–hydrogencarbonate ion equilibrium*

Worked example: Calculating the HCO_3^-/H_2CO_3 concentration ratio in a blood sample

The pH of a blood sample is measured as 7.28.

Calculate the HCO_3^-/H_2CO_3 concentration ratio in the blood sample.

(K_a for $H_2CO_3 = 7.9 \times 10^{-7}\ \text{mol dm}^{-3}$)

Step 1: Convert pH into $[H^+(aq)]$

$$[H^+(aq)] = 10^{-7.28} = 5.25 \times 10^{-8}\ \text{mol dm}^{-3}$$

Step 2: Express $[HCO_3^-]/[H_2CO_3]$ in terms of K_a and $[H^+(aq)]$

$$H_2CO_3(aq) \rightleftharpoons H^+(aq) + HCO_3^-(aq) \qquad K_a = \frac{[H^+(aq)]\,[HCO_3^-(aq)]}{[H_2CO_3(aq)]}$$

$$\therefore \frac{[HCO_3^-(aq)]}{[H_2CO_3(aq)]} = \frac{K_a}{[H^+(aq)]}$$

Step 3: Calculate the HCO_3^-/H_2CO_3 concentration ratio.

$$\frac{[HCO_3^-(aq)]}{[H_2CO_3(aq)]} = \frac{7.9 \times 10^{-7}}{5.25 \times 10^{-8}} = \frac{15}{1}$$

Summary questions

1 a Write the equilibrium for the H_2CO_3/HCO_3^- buffer system and its K_a expression. *(2 marks)*

b Explain, in terms of equilibrium, how this buffer system removes excess acid and alkali from the blood. *(4 marks)*

2 A blood sample has a HCO_3^-/H_2CO_3 ratio of 12 : 1 (K_a for $H_2CO_3 = 7.9 \times 10^{-7}\ \text{mol dm}^{-3}$) Calculate the pH of the blood sample. *(3 marks)*

3 Healthy blood has a pH range of 7.35–7.45. Calculate the range of HCO_3^-/H_2CO_3 ratios for this pH range. *(4 marks)*

21.3 Neutralisation

Specification reference: 5.1.3

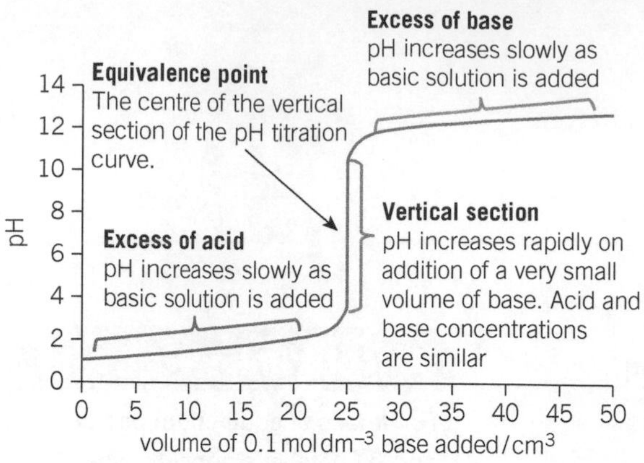

▲ **Figure 1** *Features of an acid–base titration curve*

pH titration curves

pH titration curves show the pH changes that take place during an acid–base titration. Figure 1 shows the key features of a pH titration curve for addition of an aqueous base to an aqueous acid until the base is in great excess.

Using a pH meter

A pH meter provides a convenient way of measuring pH accurately. The pH meter is connected to an electrode which is placed into a solution. The meter displays the pH reading, typically to two decimal places. A pH meter can be used to obtain the pH readings for a pH titration curve.

Synoptic link

Look back to Topic 4.2, Acid–base titrations.

Indicators

The end point

An acid–base indicator is a weak acid, HA, that has a different colour from its conjugate base, A^-.

At the **end point** of a titration,

- there are equal concentrations of HA and A^-
- the colour is between the two extreme colours.

Figure 2 shows the equilibrium for the indicator methyl orange with the colours for HA, A^- and at the end point.

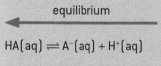

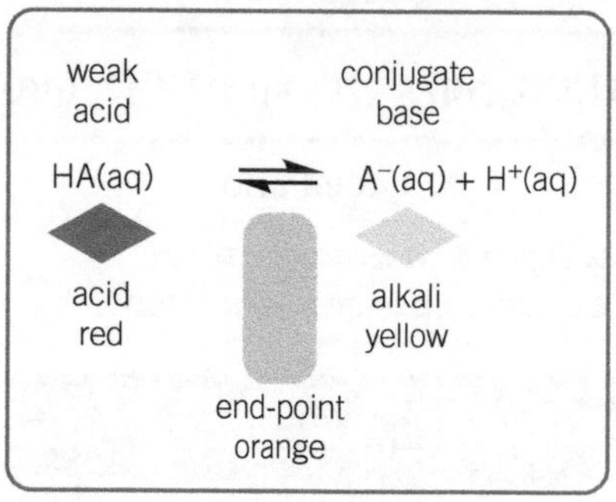

▲ **Figure 2** *The methyl orange equilibrium*

Indicator colour changes and equilibrium

An indicator changes colour in response to a shift in the equilibrium position.

Indicator colour change in alkali

On addition of methyl orange to an alkaline solution:

- $OH^-(aq)$ ions from the alkali react with $H^+(aq)$ ions in the equilibrium
- HA dissociates, shifting the equilibrium position to the right
- the indicator colour changes to yellow.

Indicator colour change in acid

On addition of methyl orange to an acid solution:

- $H^+(aq)$ ions from the acid react with $A^-(aq)$ from the indicator
- the equilibrium position shifts to the left
- the indicator changes colour to red.

Revision tip

In acid conditions, the equilibrium position shifts towards the weak acid HA.

equilibrium ←

$HA(aq) \rightleftharpoons A^-(aq) + H^+(aq)$

In basic conditions, the equilibrium position shifts towards the conjugate base, A^-.

equilibrium →

$HA(aq) \rightleftharpoons A^-(aq) + H^+(aq)$

Indicators and titration curves

Different indicators have different pK_a values and they change colour over different pH ranges. Table 1 shows examples of pH ranges for different indicators. The pH at the midpoint of the range is approximately equal to the pK_a value of the indicator.

▼ **Table 1** *pH ranges for different indicators*

Indicator	pK_a	pH range
bromophenol blue	3.9	3.6–4.6
methyl orange	3.7	3.2–4.4
metacresol purple	8.3	7.4–9.0
phenolphthalein	9.4	8.2–10.0

Synoptic link

The equilibrium shift for indicators is like the mode of action of buffer solutions. For details, see Topic 21.1, Buffer solutions.

In a titration, you must use an indicator with an end point that coincides with the vertical section of the pH titration curve (see Figure 1). pH titration curves are different for different combinations of strong and weak acids with strong and weak bases.

- Figures 3–6 show pH titration curves for these combinations and pH ranges for the indicators, methyl orange, and phenolphthalein.
- If the pH range matches the vertical section of the curve, the colour will change on addition of a very small volume (1–2 drops) and the indicator is then suitable for the titration.

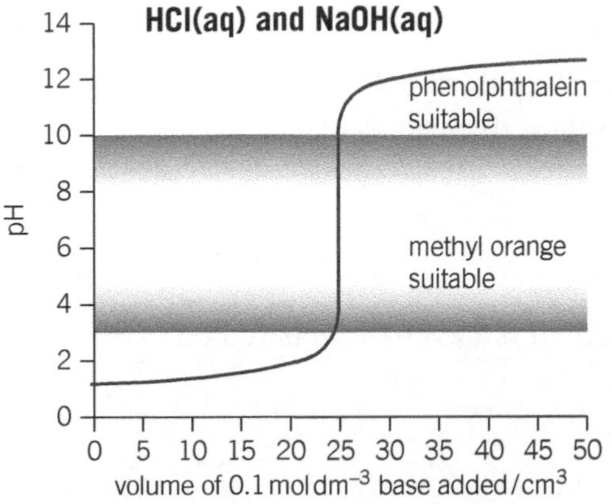

▲ **Figure 3** *Strong acid–strong base titration*

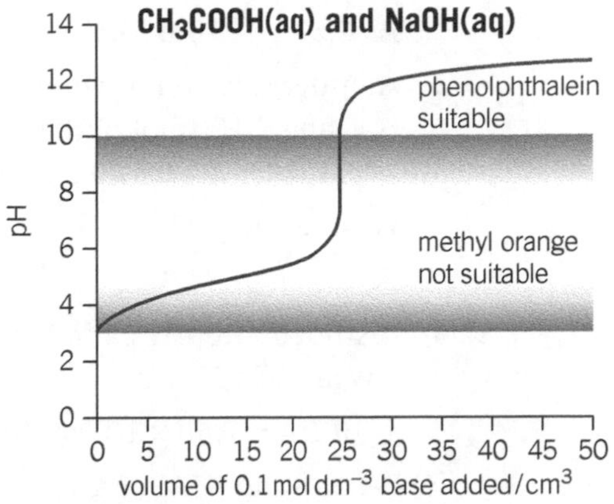

▲ **Figure 4** *Weak acid–strong base titration*

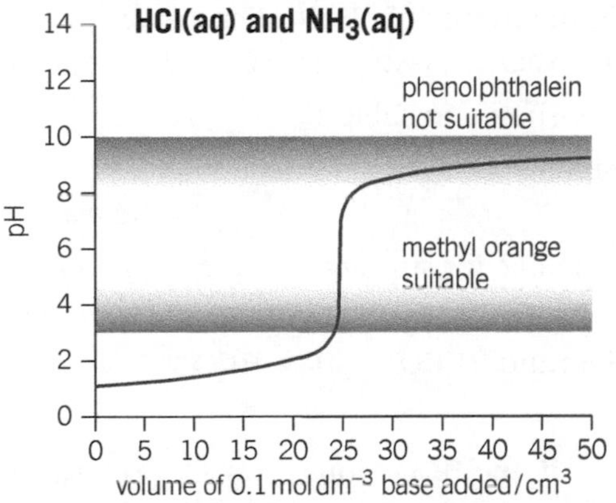

▲ **Figure 5** *Strong acid–weak base titration*

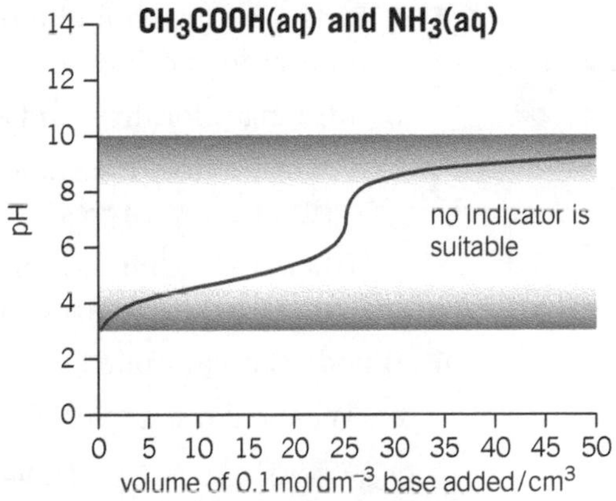

▲ **Figure 6** *Weak acid–weak base titration*

Revision link

Notice the sharp rise in pH marked by the vertical section. This is the most important part of the titration curve and shows when exact neutralisation has taken place.

Summary questions

1 a Why is the vertical section of a pH titration curve important? *(2 marks)*

b Why do different indicators have end points with different pH values? *(1 mark)*

2 a Refer to the pH ranges of the indicators in Table 1 and the pH titration curves in Figures 3–6.
Which indicators are suitable for the following titrations?
i strong acid–weak base **ii** weak acid–strong base. *(2 marks)*

b Explain why no indicator is suitable for titration of a weak acid with a weak base. *(1 mark)*

3 For the indicator bromothymol blue, HA is yellow and A^- is blue.

a Predict the colour at the end point. *(1 mark)*

b Explain in terms of equilibrium the colour change of bromothymol blue

i on addition to an alkali *(3 marks)*

ii on addition to an acid? *(3 marks)*

Chapter 21 Practice questions

1 Which mixture would **not** produce a buffer solution?

A 50 cm^3 1 mol dm^{-3} $CH_3COOH(aq)$ and 25 cm^3 1 mol dm^{-3} CH_3COONa

B 25 cm^3 1 mol dm^{-3} $CH_3COOH(aq)$ and 50 cm^3 1 mol dm^{-3} CH_3COONa

C 50 cm^3 1 mol dm^{-3} $CH_3COOH(aq)$ and 25 cm^3 1 mol dm^{-3} NaOH

D 25 cm^3 1 mol dm^{-3} $CH_3COOH(aq)$ and 50 cm^3 1 mol dm^{-3} NaOH *(1 mark)*

2 Which pK_a value of a weak acid would be most suitable to make a buffer at pH 3.8?

A 2.5 B 3.5 C 4.5 D 5.5 *(1 mark)*

3 A buffer solution of volume 1.00 dm^3 contains 0.200 mol of a weak acid, HA, and 0.100 mol of the sodium salt of the acid, NaA.
(K_a of HA = 2.51×10^{-5} mol dm^{-3})

What is the pH?

A 2.3 B 3.3 C 4.3 D 7.0 *(1 mark)*

4 A student prepares a buffer solution with a pH of 4.50 based on a weak acid HA.

K_a of HA = 1.58×10^{-5} mol dm^{-3}.

What is the [HA] : [A^-] ratio in the buffer solution?

A 1 : 2 B 2 : 1 C 4.8 : 1 D 1 : 4.8 *(1 mark)*

▼ **Table 1** *pH ranges of indicators*

Indicator	pH range
2,4-nitrophenol	2.8–4.0
cresol red	7.0–8.8
thymolphthalein	9.4–10.6

5 a What are the main differences in the shapes of pH titration curves for a strong acid–weak base and a weak acid–strong base? *(2 marks)*

b pH ranges for three indicators are shown in Table 1.

i Explain how the choice of a suitable indicator is linked to pH titration curves. *(2 marks)*

ii Select which indicator(s) are suitable for strong acid–weak base and weak acid–strong base titrations. *(2 marks)*

6 Blood pH is controlled by the equilibrium: $H_2CO_3 \rightleftharpoons H^+ + HCO_3^-$.

a What are the names for

i H_2CO_3 *(1 mark)* ii the HCO_3^- ion. *(1 mark)*

b Write the K_a expression for H_2CO_3. *(1mark)*

c Explain how the H_2CO_3/HCO_3^- buffer system controls blood pH. *(4 marks)*

7 Calculate the pH of the following buffer solutions based on HCOOH.
($K_a = 1.70 \times 10^{-4}$ mol dm^{-3})

a A buffer with concentrations 0.250 mol dm^{-3} HCOOH and 0.450 mol dm^{-3} HCOONa. *(2 marks)*

b A buffer solution prepared by mixing 25 cm^3 of 0.250 mol dm^{-3} HCOOH and 75 cm^3 0.200 mol dm^{-3} HCOONa. *(4 marks)*

c A buffer solution prepared by adding 400 cm^3 0.250 mol dm^{-3} NaOH(aq) to 600 cm^3 0.400 mol dm^{-3} HCOOH. *(4 marks)*

22.1 Lattice enthalpy

Specification reference: 5.2.1

Lattice enthalpy

Lattice enthalpy is a measure of ionic bond strength in a giant ionic lattice.

Lattice enthalpy, $\Delta_{LE}H$, is the enthalpy change that accompanies the formation of one mole of an ionic compound from its gaseous ions under standard conditions.

e.g. $\underbrace{Na^+(g) + Cl^-(g)}_{\substack{\text{gaseous} \\ \text{ions}}} \rightarrow \underbrace{NaCl(s)}_{\substack{\text{solid} \\ \text{ionic lattice}}}$ $\Delta_{LE}H^{\ominus} = -790\,kJ\,mol^{-1}$

Construction of Born–Haber cycles

Lattice enthalpy cannot be measured directly and is calculated indirectly using an energy cycle called a **Born–Haber cycle**.

Energy changes in a Born–Haber cycle

A Born–Haber cycle contains different types of energy changes, which are described below for sodium chloride.

Enthalpy change of atomisation

The standard enthalpy change of atomisation, $\Delta_{at}H^{\ominus}$, is the enthalpy change for the formation of one mole of gaseous atoms from the element in its standard state under standard conditions.

The equations show the enthalpy changes of atomisation of Na and Cl.

A $Na(s) \qquad \rightarrow Na(g)$ $\Delta_{at}H$ of sodium

B $\frac{1}{2}Cl_2(g) \quad \rightarrow Cl(g)$ $\Delta_{at}H$ of chlorine

First ionisation energy

First ionisation energy is the enthalpy change for the **removal** of one electron from each atom in one mole of gaseous atoms to form one mole of gaseous 1 + ions.

C $Na(g) \rightarrow Na^+(g) + e^-$ $\Delta_{IE}H$ of sodium

Electron affinity

First electron affinity is the enthalpy change for the **addition** of one electron to each atom in one mole of gaseous atoms to form one mole of gaseous 1− ions.

D $Cl(g) + e^- \rightarrow Cl^-(g)$ $\Delta_{EA}H$ of chlorine

Lattice enthalpy

E $Na^+(g) + Cl^-(g) \rightarrow NaCl(s)$ $\Delta_{LE}H$ of sodium chloride

    Gaseous ions → solid ionic lattice

Enthalpy change of formation

Enthalpy change of formation is the enthalpy change for the formation of one mole of a compound from its elements in their standard states under standard conditions.

F $Na(s) + \frac{1}{2}Cl_2(g) \rightarrow NaCl(s)$ $\Delta_f H$ of sodium chloride

Synoptic link

In Topic 5.2, Ionic bonding and structure, you learnt that the solid structure of an ionic compound is a giant ionic lattice.

Refer to Topic 9.1, Enthalpy changes, for important enthalpy changes.

Revision tip

Bond *forming* is an *exothermic* process.

Lattice enthalpy:
- is formation of an ionic bond from separate gaseous ions
- is an *exothermic* change.

Revision tip

$\Delta_{at}H$ applies to an element in standard state → gaseous atoms.

Synoptic link

$\Delta_{IE}H$ of an element applies to: gaseous atoms → gaseous cations (+ ions).

To revise ionisation energies, look back at Topic 7.2, Ionisation energies.

Revision tip

$\Delta_{EA}H$ of an element applies to: gaseous atoms → gaseous anions (− ions).

Contrast with $\Delta_{IE}H$: atoms → cations (+ ions).

Synoptic link

$\Delta_f H$ applies to: elements in standard state → compound in standard state.

To revise enthalpy change of formation, see Topic 9.1, Enthalpy changes.

Synoptic link

$\Delta_{LE}H$ of a compound applies to: gaseous ions → solid ionic lattice.

See earlier in this topic for details of lattice enthalpy.

Constructing a Born–Haber cycle for NaCl

In a Born–Haber cycle, energy changes connect elements in their standard states with gaseous atoms, gaseous ions, and the solid ionic lattice.

Figure 1 shows a Born–Haber cycle for NaCl based on the energy changes **A–F** from the previous section.

- The values for all energy changes except for the lattice enthalpy of NaCl have been added.

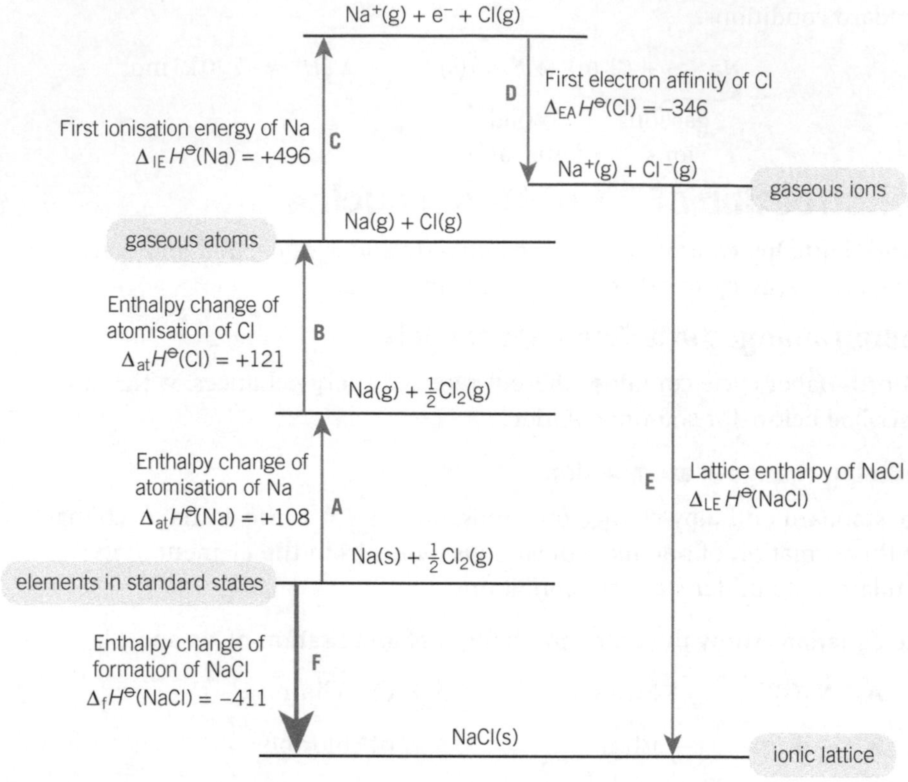

▲ **Figure 1** *Born–Haber cycle for NaCl. All enthalpy changes are in kJ mol^{-1}*

Revision tip

Look carefully at the energy changes in the Born–Haber cycle and match them to the energy changes **A–F** described in the previous section and in the Worked example below.

- Look carefully at what happens between the energy levels.
- One species changes at a time and all species present are included.

Revision tip

In Figure 1, just follow the arrows to see the two routes. Both routes start and finish at the same energy level.

- The arrows start at 'elements in standard states': Na(s) + Cl_2(g).
- The arrows finish at 'ionic lattice': NaCl(s).

Revision tip

Although this example calculates an unknown lattice enthalpy, any of the energy changes can be determined, provided that all other enthalpy changes are known.

Worked example: Calculating a lattice energy from a Born–Haber cycle

Calculate the lattice enthalpy of NaCl from the Born–Haber cycle in Figure 1.

Step 1: Use Hess' law to link all the enthalpy changes.

In Figure 1, there are two energy routes between **elements in standard states** and **ionic lattice**:

Route 1:	**A + B + C + D + E**	*clockwise from Na(s) + Cl_2(g)*
Route 2:	**F**	*anticlockwise from Na(s) + Cl_2(g)*

Using Hess's law: **A + B + C + D + E = F**

Step 2: Calculate the lattice enthalpy of NaCl (enthalpy change **E**).

Substituting the values for the enthalpy changes:

+108 + 121 + 496 + (−346) + **E** = −411

∴ $\Delta_{LE}H$ (NaCl) (**E**) = −411 − 379 = −790 kJ mol^{-1}

Other Born–Haber cycles

Group 2 compounds

Born–Haber cycles for Group 2 compounds require first and second ionisation energies. These are simple stacked one above the other in the Born–Haber cycle.

Group 2 halides have formulae such as $MgCl_2$.

In a Born–Haber cycle, you will need to form

- 2Cl(g), requiring $2 \times \Delta_{at}H(Cl) = 2 \times 121 = 242\,kJ\,mol^{-1}$
- $2Cl^-(g)$, requiring $2 \times \Delta_{EA}H(Cl) = 2 \times -346 = -692\,kJ\,mol^{-1}$

Oxides and sulfides

Born–Haber cycles for oxides require first and second electron affinities.

- The first electron affinity is exothermic.

  $O(g) + e^- \rightarrow O^-(g) \qquad \Delta_{EA1}H = -141\,kJ\,mol^{-1}$

- The second electron affinity is endothermic.

  $O^-(g) + e^- \rightarrow O^{2-}(g) \qquad \Delta_{EA2}H = +790\,kJ\,mol^{-1}$

The second electron affinity of oxygen is endothermic because:

- An electron is being gained by a $O^-(g)$ ion, which repels the negative electron (same charges).
- Energy is needed to force the negatively charged electron onto the negative ion.

Summary questions

1 a Name the enthalpy changes for the following:

- **i** $Li(s) \rightarrow Li(g)$ *(1 mark)*
- **ii** $Br(g) + e^- \rightarrow Br^-(g)$ *(1 mark)*
- **iii** $Ca^{2+}(g) + 2I^-(g) \rightarrow CaI_2(s)$ *(1 mark)*
- **iv** $Mg^+(g) \rightarrow Mg^{2+}(g) + e^-$ *(1 mark)*

2 a Write equations for the changes that accompany the following.

- **i** enthalpy change of formation of potassium bromide *(1 mark)*
- **ii** lattice enthalpy of sodium oxide *(1 mark)*
- **iii** enthalpy change of atomisation of fluorine. *(1 mark)*

b You are provided with the information:

$\Delta_{at}H^\ominus(K) = +89\,kJ\,mol^{-1}$ $\quad \Delta_{at}H^\ominus(I) = +107\,kJ\,mol^{-1}$

$\Delta_{IE}H^\ominus(K) = +419\,kJ\,mol^{-1}$ $\quad \Delta_{EA}H^\ominus(I) = -295\,kJ\,mol^{-1}$

$\Delta_f H^\ominus(KI) = -328\,kJ\,mol^{-1}$

- **i** Construct a Born–Haber cycle for KI. *(5 marks)*
- **ii** Calculate the lattice enthalpy of KI. *(2 marks)*

3 a Write the equation for the second electron affinity of sulfur. *(1 mark)*

b You are provided with the information: $\Delta_{at}H^\ominus(Mg) = +150\,kJ\,mol^{-1}$;

$\Delta_{at}H^\ominus(S) = +279\,kJ\,mol^{-1}$ $\quad \Delta_{IE1}H^\ominus(Mg) = +736\,kJ\,mol^{-1}$

$\Delta_{IE2}H^\ominus(Mg) = +1450\,kJ\,mol^{-1}$ $\quad \Delta_{EA1}H^\ominus(S) = -200\,kJ\,mol^{-1}$

$\Delta_{LE}H^\ominus(MgS) = -3299\,kJ\,mol^{-1}$ $\quad \Delta_f H^\ominus(MgS) = -346\,kJ\,mol^{-1}$

- **i** Construct a Born–Haber cycle for MgS. *(5 marks)*
- **ii** Calculate the second electron affinity of sulfur. *(2 marks)*

22.2 Enthalpy changes in solution

Specification reference: 5.2.1

Synoptic link

In Topic 5.2, Ionic bonding and structure, you learnt that ionic compounds tend to dissolve in polar solvents.

You learnt about lattice enthalpy in Topic 22.1, Lattice enthalpy.

Breaking up lattice: endothermic
$$\Delta H = + 788 \, kJ \, mol^{-1}$$

NaCl(s) $\rightleftarrows$ Na$^+$(g) + Cl$^-$(g)

$$\Delta_{LE}H = -788 \, kJ \, mol^{-1}$$
Formation of lattice: exothermic

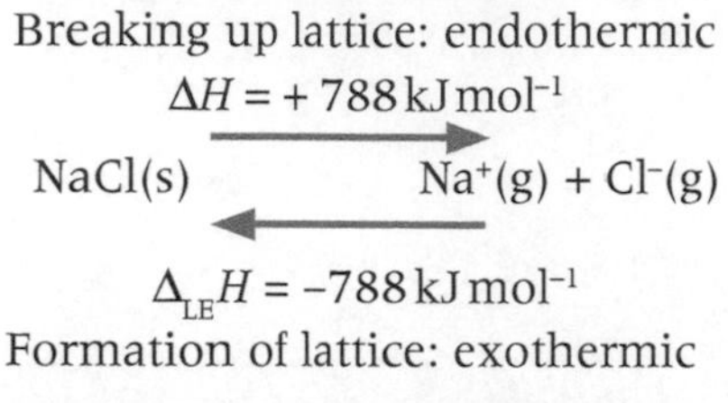

▲ **Figure 1** *Energy changes for formation and breaking up of a lattice*

Revision tip

In the equations, 'aq' represents an excess of water.

Revision tip

You are expected to know the definitions for enthalpy change of hydration and solution.

Dissolving ionic compounds in water

Two processes take place when a solid ionic compound dissolves in water:

- the ionic lattice breaks up to form gaseous ions
- the gaseous ions bond to water molecules to form aqueous ions.

Energy changes for dissolving sodium chloride in water

Lattice enthalpy

$$\textbf{A} \qquad Na^+(g) + Cl^-(g) \rightarrow NaCl(s) \qquad \Delta_{LE}H^\ominus = -790 \, kJ \, mol^{-1}$$

Energy is required to break up an ionic lattice into gaseous ions.

For dissolving, the energy change is the same as lattice energy but has the opposite sign. See Figure 1.

Enthalpy change of hydration

Enthalpy change of hydration $\Delta_{hyd}H$ is the enthalpy change when one mole of gaseous ions dissolves in water to form aqueous ions.

The equations show the enthalpy changes of hydration of Na$^+$ and Cl$^-$.

$$\textbf{B} \qquad Na^+(g) + aq \rightarrow Na^+(aq) \qquad \Delta_{hyd}H = -406 \, kJ \, mol^{-1}$$
$$\textbf{C} \qquad Cl^-(g) + aq \rightarrow Cl^-(aq) \qquad \Delta_{hyd}H = -378 \, kJ \, mol^{-1}$$

Enthalpy change of hydration is exothermic because bonds are being formed between isolated gaseous ions and water molecules.

Enthalpy change of solution

Enthalpy change of solution $\Delta_{sol}H$ is the enthalpy change when one mole of a solute dissolves in a solvent.

$$\textbf{D} \qquad Na^+Cl^-(s) + aq \rightarrow Na^+(aq) + Cl^-(aq) \qquad \Delta_{sol}H = +4 \, kJ \, mol^{-1}$$

Constructing an energy cycle

We can construct an energy cycle that connects gaseous ions with aqueous ions and the ionic lattice.

Figure 2 shows an energy cycle for the dissolving of sodium chloride, NaCl, based on the energy changes **A–D** from the previous section.

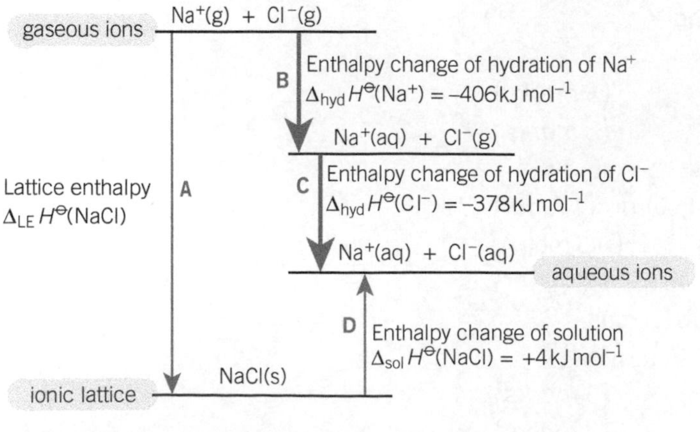

▲ **Figure 2** *Energy cycle for dissolving NaCl(s) in water. ΔH values in kJ mol^{-1}*

- The values for all energy changes have been added except the lattice enthalpy of NaCl.
- **A–D** refer to the energy changes in the previous section and in Figure 2.

Enthalpy change of solution can be exothermic or endothermic, depending on the relative sizes of the lattice enthalpy and the enthalpy changes of hydration.

- If $\Delta_{LE}H$ is less negative that $\Sigma(\Delta_{hyd}H)$, $\Delta_{sol}H$ is negative (exothermic)
- If $\Delta_{LE}H$ is more negative that $\Sigma(\Delta_{hyd}H)$, $\Delta_{sol}H$ is positive (endothermic)

Calculating the enthalpy change of hydration of Ca^{2+} ions

You are provided with the enthalpy changes in Table 1.

Construct an energy cycle and calculate the enthalpy change of hydration of Ca^{2+} ions.

Step 1: Construct the energy cycle that links gaseous ions with aqueous ions and ionic lattice.

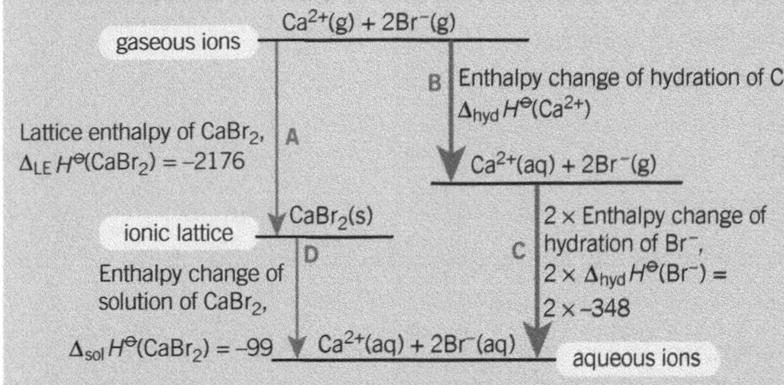

▲ **Figure 3** *Energy cycle for dissolving $CaBr_2(s)$ in water. ΔH values in kJ mol^{-1}*

Step 2: Use Hess' law to link the two routes from gaseous ions to aqueous ions.

Route 1: **A + D** *anticlockwise from $Ca^{2+}(g) + 2Br^-(g)$*

Route 2: **B + C** *clockwise from $Ca^{2+}(g) + 2Br^-(g)$*

Using Hess's law: **A + D = B + C**

Step 3: Calculate the enthalpy change of hydration of Ca^{2+} (enthalpy change **B**).

Substituting the values for the enthalpy changes:

$(-2176) + (-99) = \mathbf{B} + (2 \times -348)$

$\therefore \Delta_{hyd}H(Ca^{2+})$ (**B**) $= (-2176) + (-99) - (2 \times -348) = -1579$ kJ mol^{-1}

▼ **Table 1** *Enthalpy changes*

$\Delta_{LE}H^\ominus(CaBr_2)$ /kJ mol^{-1}	−2176
$\Delta_{hyd}H^\ominus(Br^-)$/kJ mol^{-1}	−348
$\Delta_{sol}H^\ominus(CaBr_2)$/kJ mol^{-1}	−99

Revision tip

This example is similar to the cycle shown for NaCl in Figure 2 but there is one important difference:

- Two Br^- ions are involved.
- Enthalpy changes involving $2Br^-$ need to be multiplied by 2.

Revision tip

In Figure 3, just follow the arrows to see the two routes:

- The arrows start at 'gaseous ions', $Ca^{2+}(g) + 2Br^-(g)$.
- The arrows finish at 'aqueous ions', $Ca^{2+}(aq) + 2Br^-(aq)$.

Summary questions

1 a Define enthalpy change of hydration. *(1 mark)*

b Name the enthalpy change that accompanies the following changes

i $F^-(g) + aq \rightarrow F^-(aq)$ *(1 mark)*

ii $LiF(s) \rightarrow Li^+(aq) + F^-(aq)$ *(1 mark)*

2 a Write equations for the changes that accompany the following:

i enthalpy change of hydration of calcium ions *(1 mark)*

ii lattice enthalpy of lithium bromide *(1 mark)*

iii enthalpy change of solution of calcium chloride. *(1 mark)*

b Refer back at Figure 2 and calculate the lattice enthalpy of NaCl. *(2 marks)*

3 You are provided with the information:
$\Delta_{hyd}H(Mg^{2+}) = -1926$ kJ mol^{-1}, $\Delta_{LE}H(MgI_2) = -2327$ kJ mol^{-1}, $\Delta_{sol}H(MgI_2) = -215$ kJ mol^{-1}.

a Construct an energy cycle for dissolving magnesium iodide. *(5 marks)*

b Calculate the enthalpy change of hydration of iodide ions. *(2 marks)*

22.3 Factors affecting lattice enthalpy and hydration

Specification reference: 5.2.1

Synoptic link

For details of lattice enthalpy and enthalpy change of hydration, see Topic 22.1, Lattice enthalpy, and Topic 22.2, Enthalpy changes in solution.

▼ **Table 1** *Effect of ionic size on lattice enthalpy*

Cation		Effect
Na^+ K^+ Rb^+	↓	• ionic **radius increases** • attraction between ions **decreases** • lattice enthalpy **less exothermic** • melting point **decreases**

▼ **Table 2** *Effect of ionic size on hydration enthalpy*

Cation		Effect
Na^+ K^+ Rb^+	↓	• ionic **radius increases** • attraction for H_2O **decreases** • Hydration enthalpy **less exothermic**

Revision tip

When comparing lattice enthalpy and enthalpy change of hydration don't use 'bigger' and 'smaller'.

Instead, use 'more exothermic/ more negative' or 'less exothermic/ less negative'. Enthalpy changes with negative values are difficult to compare using 'bigger and 'smaller.

Factors affecting lattice enthalpy and hydration

Lattice enthalpy and enthalpy change of hydration (hydration enthalpy) depend on: **ionic size** and **ionic charge**.

Lattice enthalpy

As **ionic size increases**:

- attraction between oppositely charged ions **decreases**
- lattice enthalpy becomes **less** exothermic
- melting point decreases as less energy is required to overcome the decreased attraction.

Effect of ionic charge on lattice enthalpy

As **ionic charge increases:**

- attraction between oppositely charged ions **increases**
- lattice enthalpy becomes **more** exothermic
- melting point increases as more energy is required to overcome the increased attraction.

Hydration enthalpy

Hydration enthalpy is affected by ionic size and charge in a similar way to lattice enthalpy.

As **ionic size decreases** and **ionic charge increases**:

- attraction between ion and water molecules **increases**
- enthalpy change of hydration becomes **more** exothermic.

Summary questions

1 State and explain the factors that affect the values of hydration enthalpies. *(2 marks)*

2 Explain the differences in lattice enthalpies and hydration enthalpies shown below:

a $\Delta_{LE}H(NaBr) = -742\ kJ\ mol^{-1}$ and $\Delta_{LE}H(KBr) = -679\ kJ\ mol^{-1}$ *(2 marks)*

b $\Delta_{hyd}H(Na^+) = -406\ kJ\ mol^{-1}$ and $\Delta_{hyd}H(Ca^{2+}) = -1579\ kJ\ mol^{-1}$ *(2 marks)*

3 When comparing the effect of ionic charge on lattice enthalpy, suggest why it is better to compare NaCl with $CaCl_2$ than NaCl with $MgCl_2$. *(2 marks)*

22.4 Entropy

Specification reference: 5.2.2

Entropy

Energy has a natural tendency to spread out rather than be concentrated in one place.

Entropy, *S*, is a measure of the dispersal of energy in a system.

The greater the entropy:

- the greater the dispersal of energy
- the greater the disorder.

Standard entropies, $S^\ominus$

Standard entropy, $S^\ominus$, is the entropy content of one mole of a substance, under standard conditions: a pressure of 100 kPa and a temperature of 298 K.

- Standard entropies have units of $J\,K^{-1}\,mol^{-1}$.
- Standard entropies are always positive.

Revision tip

At 0 K there is no energy and the standard entropy of all substances is $0\,J\,K^{-1}\,mol^{-1}$.

Above 0 K, all substances possess energy and all standard entropies have a positive sign.

Entropy change, ΔS

When a system changes to become more random:

- energy is more spread out
- the entropy change, ΔS, is positive.

When a system changes to become less random:

- energy is more concentrated
- the entropy change, ΔS, is negative.

ΔS and changes of state

As a substance changes state from solid → liquid → gas:

- particles are arranged more randomly
- energy becomes more spread out
- entropy increases and ΔS is positive.

e.g. When ice melts $H_2O(s) \rightarrow H_2O(l)$ $\Delta S = +29\,J\,K^{-1}\,mol^{-1}$

solid → liquid

When water boils $H_2O(l) \rightarrow H_2O(g)$ $\Delta S = +119\,J\,K^{-1}\,mol^{-1}$

liquid → gas

ΔS and changes in the moles of gaseous molecules

Increase in moles of gaseous molecules

An **increase** in the moles of gaseous molecules **increases** the entropy.

e.g. $\underbrace{Zn(s) + 2HCl(aq)}_{\text{0 mol of gas}} \rightarrow \underbrace{ZnCl_2(aq) + H_2(g)}_{\text{1 mol of gas}}$ $n(\text{gas}) = +1$ ΔS increases

Decrease in moles of gaseous molecules

A **decrease** in the moles of gaseous molecules **decreases** in entropy.

e.g. $\underbrace{2SO_2(g) + O_2(g)}_{\text{3 mol of gas}} \rightarrow \underbrace{2SO_3(g)}_{\text{2 mol of gas}}$ $n(\text{gas}) = -1$ ΔS decreases

Revision tip

- Production of a gas → more disorder.
- Energy is more spread out and ΔS is +ve.
- Decrease in moles of gas molecules → less disorder.
- Energy is more concentrated and ΔS is −ve.

Calculation of entropy changes, ΔS

Standard entropies can be used to calculate the entropy change of a reaction, ΔS.

$$\Delta S = \sum S^{\ominus} \text{ (products)} - \sum S^{\ominus} \text{ (reactants)} \quad (\Sigma = \text{'sum of'})$$

Synoptic link

This method is similar to calculating the enthalpy change of reaction using enthalpy changes of formation. See Topic 9.4, Hess' law and enthalpy cycles.

Worked example: Calculating an entropy change of reaction

Calculate the entropy change of reaction for the reaction below:

$$2CH_4(g) \rightarrow C_2H_2(g) + 3H_2(g)$$

You are provided with the standard entropies in Table 1.

▼ **Table 1** *Standard entropies*

Substance	$S^{\ominus}$ / J K^{-1} mol^{-1}
$CH_4(g)$	+186.3
$C_2H_2(g)$	+200.9
$H_2(g)$	+130.7

Step 1: Link the entropies with the equation for calculating ΔS.

$$\Delta S = \sum S \text{ (products)} - \sum S \text{ (reactants)}$$

$$= [S^{\ominus}(C_2H_2) + (\mathbf{3} \times S^{\ominus}(H_2))] - (\mathbf{2} \times S^{\ominus}(CH_4))$$

Step 2: Substitute in the $S^{\ominus}$ values and calculate ΔS.

$$\Delta S = [200.9 + (\mathbf{3} \times 130.7)] - (\mathbf{2} \times 186.3) = +220.4 \text{ J K}^{-1} \text{ mol}^{-1}$$

Revision tip

Check that the sign for ΔS is correct. This reaction **increases** the total moles of gaseous moles: 2 mol → 4 mol.

The final entropy change ΔS should have a **positive** sign.

Summary questions

1 State and explain whether each change would be accompanied by an increase or decrease in entropy:

a $Br_2(l) \rightarrow Br_2(g)$ *(1 mark)*

b $CuCO_3(s) + H_2SO_4(aq) \rightarrow CuSO_4(aq) + CO_2(g) + H_2O(l)$ *(1 mark)*

c $CO(g) + 2H_2(g) \rightarrow CH_3OH(g)$ *(1 mark)*

2 Use the data in Table 2 to calculate ΔS for the reactions below:

a $Fe_2O_3(s) + 3CO(g) \rightarrow 2Fe(s) + 3CO_2(g)$ *(2 marks)*

b $C_4H_{10}(g) + 6\frac{1}{2}O_2(g) \rightarrow 4CO_2(g) + 5H_2O(l)$ *(2 marks)*

▼ **Table 2** *Standard entropies*

	$Fe_2O_3(s)$	$CO(g)$	$Fe(s)$	$CO_2(g)$	$C_4H_{10}(g)$	$O_2(g)$	$H_2O(l)$
$S^{\ominus}$ / J K^{-1} mol^{-1}	+87.4	+197.7	+27.3	+213.6	+310.1	+205.0	+69.9

3 a Calculate the standard entropy of $C_8H_{18}(g)$ from the information below and in Table 2. *(2 marks)*

$C_8H_{18}(g) + 12\frac{1}{2}O_2(g) \rightarrow 8CO_2(g) + 9H_2O(l)$ $\Delta S^{\ominus}$ $-360.7 =$ J K^{-1} mol^{-1}

b Calculate the standard entropy of $H_2(g)$ from the information below. *(2 marks)*

$N_2(g) + 3H_2(g) \rightarrow 2NH_3(g)$ $\Delta S^{\ominus}$ $-198.8 =$ J K^{-1} mol^{-1}

$S^{\ominus}(N_2(g)) = +191.6$ J K^{-1} mol^{-1}; $S^{\ominus}(NH_3(g)) = +192.3$ J K^{-1} mol^{-1}

22.5 Free energy

Specification reference: 5.2.2

Free energy

Feasibility describes whether a process can take place. A reaction is 'energetically feasible' if the products have a lower overall energy than the reactants.

The **free energy change**, ΔG, is the overall energy change in a reaction.

ΔG, is made up of two types of energy:

- **The enthalpy change (ΔH)**
  This is the heat transfer between the chemical system and the surroundings.
- **The entropy change at the temperature of the reaction, $T\Delta S$**
  This is the dispersal of energy within the chemical system itself.

The Gibbs' equation

The Gibbs' equation in Figure 1 shows the relationship between ΔG, ΔH, and $T\Delta S$.

Using free energy to predict feasibility

For a reaction to be **feasible**, there must be a **decrease** in free energy:

- $\Delta G < 0$
- $\Delta H - T\Delta S < 0$

Units

- ΔG and ΔH have units of $kJ\,mol^{-1}$
- ΔS has units of $J\,mol^{-}K^{-1}$.

Using the Gibbs' equation, ΔG can be calculated from ΔH, T, and ΔS. It is essential that the units of ΔS are first converted to $kJ\,mol^{-1}\,K^{-1}$ by dividing by 1000. This then matches the kJ in ΔH.

Feasibility at different temperatures

The feasibility of a reaction depends upon the balance between ΔH and $T\Delta S$.

- At low temperatures, ΔH has a much larger magnitude that $T\Delta S$. ΔG is largely dependent on ΔH.
- As temperature increases, the $T\Delta S$ term becomes more significant. If the temperature is high enough, $T\Delta S$ may outweigh ΔH and feasibility may change.

Limitations of predictions made for feasibility

ΔG is used to predict feasibility in terms of energy, but ΔG takes no account of the rate of reaction.

The reaction between nitrogen and hydrogen to form ammonia has a negative ΔG value at 25 °C:

$$N_2(g) + 3H_2(g) \rightarrow 2NH_3(g) \qquad \Delta G = -33.0\,kJ\,mol^{-1}$$

Although ΔG is negative, this reaction does not appear to take place at 25 °C.

The reaction has a large activation energy resulting in a very slow rate.

So, although the sign of ΔG indicates the **thermodynamic** feasibility, it takes no account of the **kinetics** or rate of a reaction.

free energy change

temperature in K

$$\Delta G = \Delta H - T\Delta S$$

enthalpy change with surroundings

entropy change of system

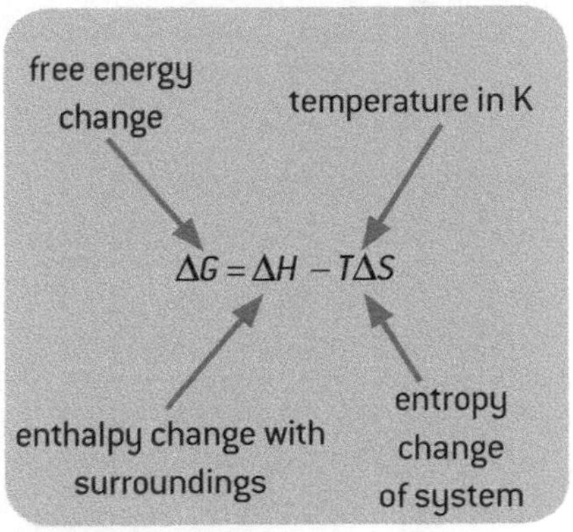

▲ **Figure 1** *The Gibbs' equation*

> **Revision tip**
> A reaction is feasible when
> - $\Delta G < 0$ or
> - $\Delta H - T\Delta S < 0$

> **Revision tip**
> If the reaction were left for long enough, it may take place but the rate may be so slow that this could take millions of years.

Synoptic link

Refer back to Topic 22.4, Entropy, for details of how to calculate an entropy change of reaction from standard entropies.

Look back to Topic 9.4, Hess' law and enthalpy changes, for calculating an enthalpy change of reaction from enthalpy changes of formation.

Revision tip

Ensure that the units are consistent.

ΔH: kJ mol^{-1}; ΔS: J K^{-1} mol^{-1}

- Convert T in °C to K by adding 273.
- Convert ΔS in J K^{-1} mol^{-1} into kJ K^{-1} mol^{-1} by ÷ 1000.

Revision tip

You need to remember to convert the ΔS value into units of kJ mol^{-1} K^{-1} to match the kJ in the ΔH value. You cannot mix units! The calculated ΔG value will have units of kJ mol^{-1}

Worked example: Determination of feasibility

Determine the feasibility of the reaction below at 25 °C.

$$N_2(g) + 3H_2(g) \rightarrow 2NH_3(g) \quad \Delta H = -92.2\ \text{kJ mol}^{-1}; \quad \Delta S = -198.8\ \text{J mol}^{-1}\text{K}^{-1}$$

Step 1: Make units consistent.

Convert T to K: $T = 273 + 25 = 298$ K

Convert ΔS to kJ mol^{-1} K^{-1}: $\Delta S = \dfrac{-198.8}{1000} = -0.1988$ kJ mol^{-1} K^{-1}

Step 2: Use the Gibbs' equation to calculate ΔG at 25°C.

$\Delta G = -92.2 - 298 \times -0.1988 = -33.0$ kJ mol^{-1}

As $\Delta G < 0$, the reaction is feasible at 25 °C.

Worked example: Calculating the minimum temperature for feasibility

Calculate the minimum temperature, in °C, for the reaction below to be feasible.

$$Cr_2O_3(s) + 3C(s) \rightarrow 2Cr(s) + 3CO(g) \quad \Delta H = +808.2\ \text{kJ mol}^{-1}; \quad \Delta S = +542.1\ \text{J mol}^{-1}\text{K}^{-1}$$

Step 1: Make units consistent.

Convert ΔS to kJ mol^{-1} K^{-1}: $\Delta S = \dfrac{+542.1}{1000} = +0.5421$ kJ mol^{-1} K^{-1}

Step 2: Determine the relationship for minimum temperature.

At the minimum temperature for feasibility, $\Delta G = \Delta H - T\Delta S = 0$

Rearranging $\Delta H - T\Delta S = 0$: $T = \dfrac{\Delta H}{\Delta S}$

Step 3: Calculate the minimum temperature

Substitute ΔH and ΔS: $T = \dfrac{808.2}{0.5421} = 1491$ K

Convert K to °C: $T = 1491 - 273 = 1218$ °C

Summary questions

1 **a** What are the conditions for feasibility? *(1 mark)*

b What are the limitations of using ΔG to predict feasibility? *(1 mark)*

2 You are provided with the following information:

$$Li_2CO_3(s) \rightarrow Li_2O(s) + CO_2(g) \quad \Delta H = +224.5.0\ \text{kJ mol}^{-1}$$
$$\Delta S = +160.8\ \text{J mol}^{-1}\text{K}^{-1}$$

a Why does the reaction have a positive ΔS value? *(1 mark)*

b Show whether the reaction is feasible at 25 °C. *(2 marks)*

c Calculate the minimum temperature, in °C, for the reaction to be feasible. *(2 marks)*

3 You are provided with the following information:

$$Pb(NO_3)_2(s) \rightarrow PbO(s) + 2NO_2(g) + \tfrac{1}{2}O_2(g)$$

	$Pb(NO_3)_2(s)$	$PbO(s)$	$NO_2(s)$	$O_2(g)$
$S^{\ominus}$ / J mol^{-1} K^{-1}	+213.0	+68.7	+240.0	+205.0
$\Delta_f H^{\ominus}$ / kJ mol^{-1}	−451.9	−217.3	+33.2	0

a Calculate ΔS and ΔH for this reaction. *(2 marks)*

b Show whether the reaction is feasible at 25 °C. *(2 marks)*

c Calculate the minimum temperature, in °C, for this reaction to take place. *(2 marks)*

Chapter 22 Practice questions

1 Predict which compound would have the most exothermic lattice enthalpy.

A KI **B** NaI **C** CaI_2 **D** MgI_2 *(1 mark)*

2 Which enthalpy change is **not** required in an energy cycle to determine the enthalpy change of hydration of bromide ions using calcium bromide?

A The enthalpy change of formation of calcium bromide.

B The lattice enthalpy of calcium bromide.

C The enthalpy change of hydration of Ca^{2+} ions.

D The enthalpy change of solution of calcium bromide. *(1 mark)*

3 Which reaction would be expected to have a negative entropy change of reaction?

A $2HgO(s) \rightarrow 2Hg(l) + O_2(g)$

B $H_2O(l) \rightarrow H_2O(g)$

C $CH_4(g) + H_2O(g) \rightarrow 3H_2(g) + CO(g)$

D $N_2(g) + 3H_2(g) \rightarrow 2NH_3(g)$ *(1 mark)*

4 Energy changes for the reaction of nitrogen and fluorine are shown below.

$N_2(g) + 3F_2(g) \rightarrow 2NF_3(g)$ $\Delta H = -250\,kJ\,mol^{-1}$ $\Delta S = -279\,J\,mol^{-1}\,K^{-1}$

Which condition for feasibility is correct?

A Feasible at high temperatures only.

B Feasible at low temperatures only.

C Feasible at all temperatures.

D Not feasible at any temperature. *(1 mark)*

5 This question is about the lattice enthalpy of calcium fluoride.

a **i** What is meant by the term lattice enthalpy? *(1 mark)*

ii Write an equation for the lattice enthalpy of calcium fluoride. *(1 mark)*

b You are provided with the information:

$\Delta_{at}H^{\ominus}$ (Ca) = +178 kJ mol^{-1}; $\Delta_{at}H$ (F) = +79 kJ mol^{-1};

$\Delta_{IE1}H^{\ominus}$ (Ca) = +590 kJ mol^{-1}; $\Delta_{IE2}H$ (Ca) = +1145 kJ mol^{-1};

$\Delta_{EA}H^{\ominus}$ (F) = −328 kJ mol^{-1}; $\Delta_f H^{\ominus}$ (CaF_2) = −1220 kJ mol^{-1}

i Construct a Born–Haber cycle for CaF_2. *(5 marks)*

ii Calculate the lattice enthalpy of CaF_2. *(2 marks)*

c State and explain the factors that affect the values of lattice enthalpies. *(3 marks)*

6 You are provided with the following information:

$Fe_2O_3(s) + 3C(s) \rightarrow 2Fe(s) + 3CO(g)$

	$Fe_2O_3(s)$	C(s)	Fe(s)	CO(g)
$S^{\ominus}$ / J mol^{-1} K^{-1}	+87.4	+5.7	+27.3	+197.6
$\Delta_f H^{\ominus}$ / kJ mol^{-1}	−824.2	0	0	−110.5

a Calculate ΔS and ΔH for this reaction. *(2 marks)*

b Show whether the reaction is feasible at 25 °C. *(2 marks)*

c Calculate the minimum temperature, in °C, for this reaction to take place. *(2 marks)*

23.1 Redox reactions

Specification reference: 5.2.3

Synoptic link

For details of oxidation number rules, oxidation, and reduction, see Topic 4.3, Redox.

Revision tip

- The oxidising agent contains the atom that is reduced.
- The reducing agent contains the atom that oxidised.

Oxidising and reducing agents

A redox reaction always has an **oxidising agent** and a **reducing agent**.

The oxidising agent:

- takes electrons away from the atom that is oxidised
- contains the atom that is reduced.

The reducing agent:

- adds electrons to the atom that is reduced.
- contains the atom that is oxidised.

e.g.

$MnO_2(s) + 4HCl(aq) \rightarrow MnCl_2(aq) + Cl_2(g) + 2H_2O$

+4		+2		Mn in MnO_2 reduced
	−1		0	Cl in HCl oxidised

MnO_2 is the oxidising agent:

- MnO_2 has oxidised Cl: −1 in HCl → 0 in Cl_2.

HCl is the reducing agent:

- HCl has reduced Mn: +4 in MnO_2 → +2 in $MnCl_2$.

Synoptic link

In Topic 4.3, Redox, you learnt that the number of electrons lost during oxidation must equal the number of electrons gained during reduction.

Redox equations from half-equations

A redox equation can be written from the half-equations for reduction and oxidation.

In a balanced equation for a redox reaction:

- number of electrons lost = number of electrons gained.

Worked example: Redox equations from half-equations

Two half-equations are shown below.

Write the overall equation for the redox reaction between hydrogen peroxide, H_2O_2, and acidified manganate(VII), H^+/MnO_4^-.

Oxidation: $H_2O_2(aq) \rightarrow O_2(g) + 2H^+(aq) + 2e^-$

Reduction: $MnO_4^-(aq) + 8H^+(aq) + 5e^- \rightarrow Mn^{2+}(aq) + 4H_2O(l)$

Step 1: Balance the electrons.

Multiply the half-equations throughout to get the same number of electrons.

Oxidation × **5:** $5H_2O_2(aq) \rightarrow 5O_2(g) + 10H^+(aq) + \mathbf{10e^-}$

Reduction × **2:** $2MnO_4^-(aq) + 16H^+(aq) + \mathbf{10e^-} \rightarrow 2Mn^{2+}(aq) + 8H_2O(l)$

Step 2: Cancel electrons and any common species on both sides.

Oxidation: $5H_2O_2(aq) \rightarrow 5O_2(g) + \cancel{10H^+(aq)} + \cancel{10e^-}$

Reduction: $2MnO_4^-(aq) + \cancel{16H^+}(aq) + \cancel{10e^-} \rightarrow 2Mn^{2+}(aq) + 8H_2O(l)$
$6H^+$

Step 3: Write the overall equation by combining the two half-equations.

Overall equation: $5H_2O_2(aq) + 2MnO_4^-(aq) + 6H^+ \rightarrow 5O_2(g) + 2Mn^{2+}(aq) + 8H_2O(l)$

Revision tip

In the Worked example, the oxidation half-equation is multiplied by × 5 to give $10e^-$.

The reduction half-equation is multiplied by × 2 to also give $10e^-$.

Revision tip

Cancel the $10e^-$ in both equations, together with any common species on both sides.

Revision tip

Check that charges balance and that all formulae are correct.

Redox equations from oxidation numbers

A redox equation can be written from oxidation numbers.

In a balanced equation for a redox reaction:

- increase in oxidation number = decrease in oxidation number

Synoptic link

You learnt in Topic 4.3, Redox, that the total increase in oxidation number during oxidation must equal the total decrease in oxidation number during reduction.

Worked example: Redox equations from oxidation numbers

Zinc reacts with VO_2^+ ions in acid, H^+, to form Zn^{2+} ions, V^{2+} ions, and water, H_2O.

Construct the overall equation for this redox reaction.

Step 1: Summarise the information provided.

$$Zn + VO_2^+ + H^+ \rightarrow Zn^{2+} + V^{2+} + H_2O$$

Step 2: Assign oxidation numbers to the atoms that change oxidation number.

$$Zn + VO_2^+ + H^+ \rightarrow Zn^{2+} + V^{2+} + H_2O$$

Zn	0		+2		oxidation number change: +2
V		+5		+2	oxidation number change: −3

Step 3: Balance the oxidation number changes

$$3Zn + 2VO_2^+ + H^+ \rightarrow 3Zn^{2+} + 2V^{2+} + H_2O$$

Zn	3 × 0		3 × +2		total increase of **+6**
V		2 × +5		2 × +2	total decrease of **−6**

Step 4: Balance any remaining atoms and cancel any common species.

$$3Zn + 2VO_2^+ + 8H^+ \rightarrow 3Zn^{2+} + 2V^{2+} + 4H_2O$$

Revision tip

Zn and Zn^{2+} are multiplied by × 3 to give a total increase of +6

VO_2^+ and V^{2+} are multiplied by × 2 to give a decrease of −6.

Revision tip

Check that the charges balance and that all formulae are correct.

Interpretation and prediction of redox reactions

You might not know all species that are involved in the reaction and you might need to predict any missing reactants or products. In aqueous redox reactions, H_2O is often formed. Other likely reactants or products are H^+ and OH^- ions, depending on the conditions used.

Revision tip

Always make sensible predictions – if your predicted formula looks strange, you have probably made up chemical species that don't exist.

Summary questions

1 Identify the changes in oxidation number, the oxidising agent and the reducing agent:

a $2Al + 3CuSO_4 \rightarrow Al_2(SO_4)_3 + 3Cu$ *(3 marks)*

b $2HBr + H_2SO_4 \rightarrow Br_2 + SO_2 + 2H_2O$ *(3 marks)*

2 Construct the overall equation from the following half-equations.

a $2I^- \rightarrow I_2 + 2e^-$;
$Cr_2O_7^{2-} + 14H^+ + 6e^- \rightarrow 2Cr^{3+} + 7H_2O$ *(2 marks)*

b $H_2S \rightarrow 2H^+ + S + 2e^-$;
$MnO_4^- + 8H^+ + 5e^- \rightarrow Mn^{2+} + 4H_2O$ *(2 marks)*

3 a Using oxidation numbers, balance the following equations.

i $MnO_4^- + H^+ + Cl^- \rightarrow Mn^{2+} + H_2O + Cl_2$ *(2 marks)*

ii $VO_3^- + SO_2 + H^+ \rightarrow VO^{2+} + SO_4^{2-} + H_2O$ *(2 marks)*

b Sn reacts with HNO_3 to form SnO_2, NO_2, and one other product. Write the balanced equation. *(2 marks)*

23.2 Manganate(VII) redox titrations

Specification reference: 5.2.3

Revision tip

For redox titrations, the procedures and analysis are very similar to acid–base titrations.

Redox titrations

The **titration** technique measures the volume of one solution that reacts exactly with a volume of another solution.

In a redox titration, a solution of a **reducing agent** is titrated with a solution of an **oxidising agent**.

Manganate(VII) redox titrations

Acidified manganate(VII) is an oxidising agent and manganate(VII) titrations are used for analysing reducing agents.

A standard solution of potassium manganate(VII), $KMnO_4$, is added to the burette.

- A solution of the reducing agent is pipetted into the conical flask. Dilute sulfuric acid is added to supply H^+(aq) ions for the redox reaction.
- $KMnO_4$(aq) is added from the burette to the solution in the conical flask. The $KMnO_4$(aq) reacts with the reducing agent and its deep purple colour is decolourised.
- At the end-point, the solution changes from colourless to the first permanent pink colour. The pink colour indicates the first trace of an excess of MnO_4^-(aq) ions.

  The titration is self-indicating – no indicator is needed.

Synoptic link

For the preparation of standard solutions, and carrying out a titration, see Topic 4.2, Acid–base titrations.

Other redox titrations

The principles for redox titrations can be extended to the analysis of many different substances.

- Manganate(VII) titrations can be used to analyse a reducing agent that reduces MnO_4^- to Mn^{2+}.
- $KMnO_4$ can be replaced with other oxidising agents, such as acidified dichromate(VI), $H^+/Cr_2O_7^{2-}$.

The procedures and calculations are similar to those for iron(II)–manganate(VII) titrations.

Modifications may be needed depending on

- the oxidising agent used for the titration
- the reducing agent being analysed
- the colour change.

Iron(II)–manganate(VII) titrations

Manganate(VII) titrations can be used to analyse an iron(II) compound (the reducing agent).

The half-equations and full equation for the reaction of acidified manganate(VII) ions with iron(II) ions are shown below.

Reduction: $MnO_4^-(aq) + 8H^+(aq) + 5e^- \rightarrow Mn^{2+}(aq) + 4H_2O(l)$

Oxidation: $Fe^{2+}(aq) \rightarrow Fe^{3+}(aq) + e^-$

Overall: $MnO_4^-(aq) + 8H^+(aq) + 5Fe^{2+}(aq) \rightarrow Mn^{2+}(aq) + 5Fe^{3+}(aq) + 4H_2O(l)$

Synoptic link

For details of how to write an overall equation from two half-equations, see Topic 23.1, Redox reactions. You should be able to combine these two half-equations to give the overall equation shown.

Calculations for iron(II)–manganate(VII) titrations

Worked example: The percentage of iron in an iron ore

A metal ore contains iron in its +2 oxidation state.

- A 6.82 g sample of the ore is dissolved in dilute sulfuric acid and the resulting solution is made up to 250.0 cm^3.
- 25.0 cm^3 of this solution is titrated against 0.0200 mol dm^{-3} $KMnO_4$
  The mean titre of $KMnO_4$(aq) is 23.80 cm^3.

Calculate the percentage by mass of iron(II) in the ore sample.

Step 1: Calculate the amount of MnO_4^- that reacted.

$$\text{Number of moles, } n(MnO_4^-) = c \times \frac{V}{1000} = 0.0200 \times \frac{23.80}{1000} = \mathbf{4.76 \times 10^{-4}\ mol}$$

Step 2: Determine the amount of Fe^{2+} that reacted.

$$MnO_4^-(aq) + 8H^+(aq) + 5Fe^{2+}(aq) \rightarrow Mn^{2+}(aq) + 5Fe^{3+}(aq) + 4H_2O(l)$$

From the equation, 1 mol MnO_4^- reacts with 5 mol Fe^{2+}.

∴ Number of moles, $n(Fe^{2+}) = \mathbf{5} \times n(MnO_4^-) = \mathbf{5} \times 4.76 \times 10^{-4} = 2.38 \times 10^{-3}$ mol

Step 3: Work out the unknown information. There are several stages.

1 Scale up to find the amount of Fe^{2+} in the 250.0 cm^3 solution that you prepared.

$n(Fe^{2+})$ in 25.00 cm^3 used in the titration = 2.38×10^{-3} mol

$n(Fe^{2+})$ in 250.0 cm^3 solution = $2.38 \times 10^{-3} \times \mathbf{10} = \mathbf{2.38 \times 10^{-2}}$ mol

2 Find the mass of Fe^{2+} in the impure sample.

mass m of $Fe^{2+} = n \times M = \mathbf{2.38 \times 10^{-2}} \times 55.8 = \mathbf{1.328\,04\ g}$

3 Find the percentage, by mass, of Fe^{2+} in the ore sample.

$$\text{Percentage} = \frac{\text{mass of } Fe^{2+}}{\text{mass of ore}} \times 100$$

$$= \frac{\mathbf{1.328\,04}}{\mathbf{6.82}} \times 100 = \mathbf{19.5\,\%}$$

Synoptic link

The results from redox titrations are analysed in a similar way to acid–base titrations. See Topic 4.2, Acid–base titrations, for further details.

Revision tip

You need to use the mean titre V and the concentration c of MnO_4^-.

Revision tip

You need to use the equation and the number of moles of MnO_4^-.

Revision tip

The unrounded value (1.328 04) should be kept in the calculator and any rounding carried out once, for the final answer.

This always gives a more accurate final answer than rounding intermediate values.

Revision tip

In this analysis, the most appropriate number of significant figures (s.f.) is 3 s.f.

3 s.f. is the lowest number of significant figures in the quantities used for the titration – for the mass of the ore as 6.82 g.

It is important to round to 3 significant figures **only** for your final answer or you will introduce rounding errors in any intermediate values.

Summary questions

1 a How is the end point detected in a manganate(VII) titration? *(1 mark)*

b Why is dilute sulfuric acid added in manganate(VII) titrations? *(1 mark)*

2 a What are the oxidation number changes in an Fe^{2+}/MnO_4^- titration? *(2 marks)*

b 24.80 cm^3 of 0.0250 mol dm^{-3} $KMnO_4$ reacts with 25.0 cm^3 of $FeSO_4$(aq).
What is the concentration of the $FeSO_4$(aq)? *(3 marks)*

3 7.18 g of an impure sample of $FeSO_4 \bullet 7H_2O$ is dissolved in water and made up to 250.0 cm^3.
25.0 cm^3 samples of this solution are acidified with H_2SO_4(aq) and titrated against 0.0200 mol dm^{-3} $KMnO_4$.
The mean titre of $KMnO_4$ is 21.80 cm^3.

Calculate the percentage purity of the impure $FeSO_4 \bullet 7H_2O$ to an appropriate number of significant figures. *(5 marks)*

23.3 Iodine/thiosulfate redox titrations

Specification reference: 5.2.3

Iodine/thiosulfate redox titrations

Iodine/thiosulfate titrations are used to analyse oxidising agents.

- The oxidising agent is first reacted with an excess of iodide ions. Iodide ions are oxidised to iodine:

$$2I^-(aq) \rightarrow I_2(aq) + 2e^-$$

- The iodine generated is then titrated with thiosulfate ions, $S_2O_3^{2-}(aq)$. Thiosulfate ions, $S_2O_3^{2-}(aq)$, are oxidised and iodine, $I_2(aq)$, is reduced to iodide, $I^-(aq)$.

Oxidation: $2S_2O_3^{2-}(aq) \rightarrow S_4O_6^{2-}(aq) + 2e^-$

Reduction: $I_2(aq) + 2e^- \rightarrow 2I^-(aq)$

Overall: $2S_2O_3^{2-}(aq) + I_2(aq) \rightarrow 2I^-(aq) + S_4O_6^{2-}(aq)$

Synoptic link

For details of how to write an overall equation from two half-equations, see Topic 23.1, Redox reactions.

Carrying out the titration

An iodine/thiosulfate titration is carried out in a similar way to other titrations, but there are some key differences.

The key steps are outlined below.

- A standard solution of sodium thiosulfate, $Na_2S_2O_3(aq)$, is added to the burette.
- A solution of the oxidising agent is pipetted into the conical flask. An excess of aqueous potassium iodide, KI(aq), is added.

  The oxidising agent reacts with iodide ions to form iodine, which turns the solution a yellow-brown colour.
- $Na_2S_2O_3(aq)$ is added from the burette to the solution in the conical flask.
- As the end point approaches, the yellow-brown iodine colour fades to become a pale straw colour. Starch indicator is then added, forming a deep blue-black colour.
- At the end point, the colour changes from blue-black to colourless. This indicates that all $I_2(aq)$ has reacted.

Synoptic link

For the preparation of standard solutions, and carrying out a titration, see Topic 4.2, Acid–base titrations.

Revision tip

The colour change with starch added is much easier to see and gives a more accurate end-point.

Calculations for iodine/thiosulfate titrations

Iodine/thiosulfate titrations can be used to analyse copper(II) ions (the oxidising agent).

When an excess of KI(aq) is added to a solution containing $Cu^{2+}(aq)$ ions:

- $Cu^{2+}(aq)$ ions are reduced by $I^-(aq)$ to form a white precipitate of copper(I) iodide, CuI(s).
- $I^-(aq)$ ions are oxidised to form a yellow-brown solution of $I_2(aq)$:

$$2Cu^{2+}(aq) + 4I^-(aq) \rightarrow 2CuI(s) + I_2(aq)$$

Reduction: +2 → +1

Oxidation: −1 → 0

- The $I_2(aq)$ in the mixture is then titrated with a standard solution of sodium thiosulfate:

$$2S_2O_3^{2-}(aq) + I_2(aq) \rightarrow 2I^-(aq) + S_4O_6^{2-}(aq)$$

2 mol Cu^{2+} produces 1 mol I_2 which reacts with 2 mol $S_2O_3^{2-}$

∴ 1 mol Cu^{2+} is equivalent to 1 mol $S_2O_3^{2-}$

Synoptic link

The analysis of results is essentially the same method as used for acid–base titrations and manganate(VII) titrations.

See Topic 4.2, Acid base titrations, and Topic 23.2, Manganate(VII) redox titrations, for further details.

Worked example: Analysing the composition of brass

Brass is an alloy of copper and zinc.

- A 0.500 g sample of brass is reacted with concentrated nitric acid to form a solution containing Cu^{2+} and Zn^{2+} ions. The solution is then neutralised.
- Excess KI(aq) is added. Cu^{2+}(aq) ions oxidise I^-(aq) ions to I_2(aq).
- The iodine is titrated with 0.200 mol dm^{-3} $Na_2S_2O_3$ and 25.20 cm^3 are required to reach the end point.

Calculate the percentage by mass of copper and zinc in the brass.

Step 1: Calculate the amount of $S_2O_3^{2-}$ that reacted in the titration.

$$n(S_2O_3^{2-}) = c \times \frac{V}{1000} = 0.200 \times \frac{25.20}{1000} = \mathbf{5.04 \times 10^{-3}\ mol}$$

Step 2: Determine the amount of Cu^{2+} that reacted.

2 mol Cu^{2+} produces 1 mol I_2 which reacts with 2 mol $S_2O_3^{2-}$

$\therefore n(Cu^{2+}) = n(S_2O_3^{2-}) = \mathbf{5.04 \times 10^{-3}}$ mol

Step 3: Work out the unknown information: the percentages of Cu and Zn.

5.04×10^{-3} mol of Cu^{2+} has a mass of **$5.04 \times 10^{-3} \times 63.5 = 0.320$ g**

Mass of Zn = 0.500 − 0.320 = 0.180 g

$$\%\text{ composition by mass of Cu} = \frac{0.320}{0.500} \times 100 = 64.0\%$$

$$\%\text{ composition by mass of Zn} = \frac{0.180}{0.500} \times 100 = 36.0\%$$

Revision tip

You need to use the mean titre V and the concentration c of $S_2O_3^{2-}$.

Revision tip

See details on previous page to check why 1 mol Cu^{2+} mol is equivalent to 1 mol $S_2O_3^{2-}$.

Other iodine/thiosulfate redox titrations

Iodine/thiosulfate redox titrations can be used to analyse many oxidising agents. The procedures and calculations are similar to those in the example with copper, although some modifications may be needed depending on the oxidising agent being analysed.

Summary questions

1 a How is the end point detected in an iodine/thiosulfate titration? *(2 marks)*

b Explain the role of excess KI(aq) in the analysis of copper(II) in an I_2/ $S_2O_3^{2-}$ titration. *(2 marks)*

2 a What are the oxidation number changes in an I_2/ $S_2O_3^{2-}$ titration? (*Note:* One oxidation number is a fraction.) *(2 marks)*

b 22.40 cm^3 of 0. 0150 mol dm^{-3} $Na_2S_2O_3$ reacts with 25.0 cm^3 of $CuSO_4$(aq) to which an excess of KI(aq) has been added. What is the concentration of the $CuSO_4$(aq)? *(3 marks)*

3 1.043 g of $CuCl_2 \cdot xH_2O$ is dissolved in water and the solution made up to 250.0 cm^3.
25.0 cm^3 samples of this solution are mixed with an excess of KI(aq) and titrated against 0.0240 mol dm^{-3} $Na_2S_2O_3$. The mean titre is 25.50 cm^3.
Determine the molar mass of the copper(II) salt, the value of x, and suggest the formula of the salt. *(5 marks)*

23.4 Electrode potentials

Specification reference: 5.2.3

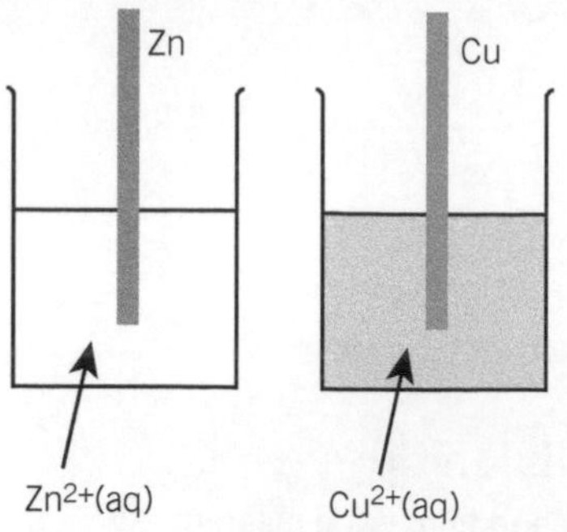

▲ **Figure 1** *$Zn^{2+}(aq) | Zn(s)$ and $Cu^{2+}(aq) | Cu(s)$ half-cells: These half-cells are based on a metal and a metal ion.*

$$Zn^{2}(aq) + 2e^{-} \rightleftharpoons Zn(s)$$

reduction (forward) / oxidation (reverse)

▲ **Figure 2** *Equilibrium in a $Zn^{2+}(aq) | Zn(aq)$ half-cell*

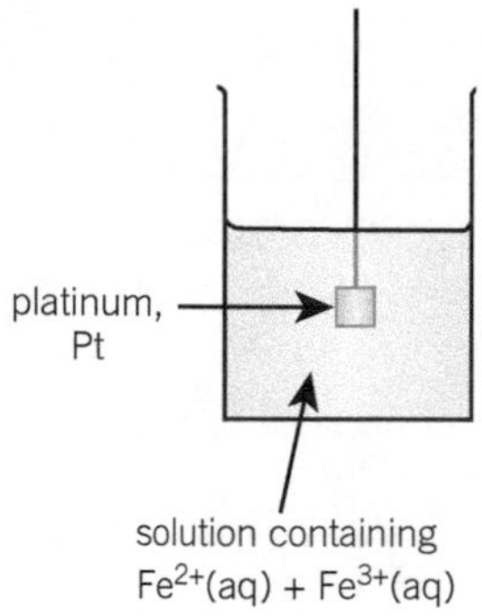

▲ **Figure 3** *$Fe^{3+}(aq)$, $Fe^{3+}(aq)$ half-cell: A half-cell based on ions of the same element in different oxidation states*

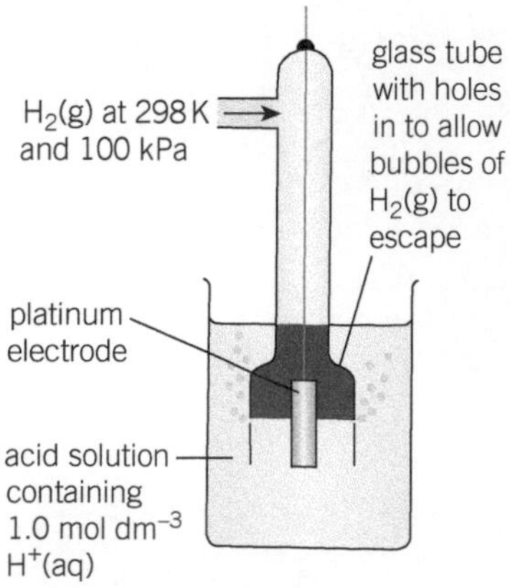

▲ **Figure 4** *Standard hydrogen electrode*

Standard electrode potential

Half-cells

A half-cell contains the chemical species present in a redox half-equation.

Metal/metal ion half-cells

This half-cell is a metal rod dipped into a solution of its metal ion (see Figure 1).

An equilibrium is set up between the metal ion and the metal. The equilibrium in a half-cell is written so that the forward reaction shows reduction and the reverse reaction shows oxidation (Figure 2).

Ion/ion half-cells

An ion/ion half-cell is a solution containing ions of the same element in different oxidation states $Fe^{2+}(aq)$ and $Fe^{3+}(aq)$.

The redox equilibrium is: $Fe^{3+}(aq) + e^{-} \rightleftharpoons Fe^{2+}(aq)$

An inert platinum electrode is used to transport electrons into or out of the half-cell (see Figure 3).

Electrode potential

Electrode potential, *E*, is the tendency for electrons to be gained and for reduction to take place in a half-cell.

Standard electrode potential

When two half-cells are connected to one another, a voltage sets up. The size of the voltage depends on the electrode potentials of each half-cell.

Standard electrode potentials are measured against a standard hydrogen electrode (Figure 4).

The **standard electrode potential**, $E^{\ominus}$, is the e.m.f. of a half-cell connected to a standard hydrogen half-cell under standard conditions of 298 K, solution concentrations of 1 mol dm^{-3}, and a pressure of 100 kPa.

The standard electrode potential of a standard hydrogen electrode is exactly 0 V.

The sign of a standard electrode potential shows:

- the polarity (+ or −) of the half-cell connected to the standard hydrogen electrode
- the relative tendency to be reduced and gain electrons compared with the hydrogen half-cell.

Measuring a standard electrode potential

To measure a standard electrode potential, the half-cell is connected to a standard hydrogen electrode.

- The two electrodes are connected by a wire to allow a controlled flow of electrons through a voltmeter.
- The two solutions are connected by a salt bridge, which allows ions to flow.
- The salt bridge contains a concentrated solution of an electrolyte that does not react with either solution, e.g. a strip of filter paper soaked in aqueous potassium nitrate, $KNO_3(aq)$.

Figure 5 shows how the standard electrode potential of a copper half-cell can be measured.

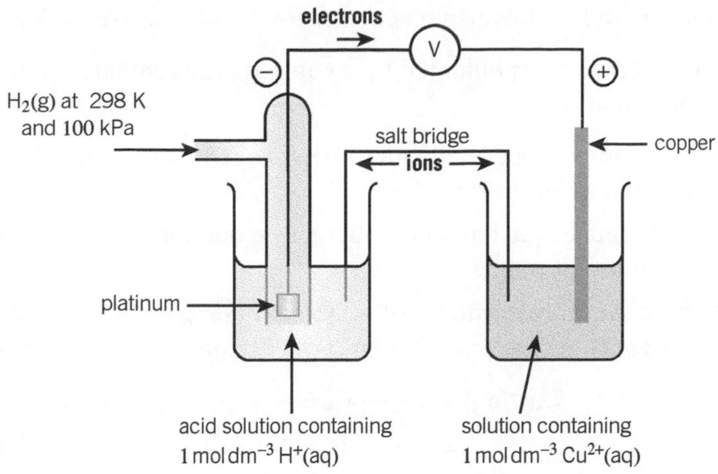

▲ **Figure 5** *Measuring a standard electrode potential*

▼ **Table 1** *Standard electrode potentials*

Redox system	$E^\ominus$ / V
$Mg^{2+}(aq) + 2e^- \rightleftharpoons Mg(s)$	−2.37
$Al^{3+}(aq) + 3e^- \rightleftharpoons Al(s)$	−1.66
$Zn^{2+}(aq) + 2e^- \rightleftharpoons Zn(s)$	−0.76
$Fe^{2+}(aq) + 2e^- \rightleftharpoons Fe(s)$	−0.44
$2H^+(aq) + 2e^- \rightleftharpoons H_2(g)$	0.00
$Cu^{2+}(aq) + 2e^- \rightleftharpoons Cu(s)$	+0.34
$I_2(aq) + 2e^- \rightleftharpoons 2I^-(aq)$	+0.54
$Fe^{3+}(aq) + e^- \rightleftharpoons Fe^{2+}(aq)$	+0.77
$Ag^+(aq) + e^- \rightleftharpoons Ag(s)$	+0.80
$Cl_2(g) + 2e^- \rightleftharpoons 2Cl^-(aq)$	+1.36

Tables of standard electrode potentials

Standard electrode potentials are listed in the data table. Table 1 shows examples of standard electrode potentials, sorted in order, with the most negative value at the top. The forward reactions show reduction.

The more negative the $E^\ominus$ value:

- the greater the tendency to lose electrons and undergo oxidation.
- the less the tendency to gain electrons and undergo reduction.

Metals tend to have negative $E^\ominus$ values and lose electrons.
Non-metals tend to have positive $E^\ominus$ values and gain electrons.

reduction

$Mg^{2}(aq) + 2e^- \rightleftharpoons Mg(s)\ E = -2.37$ V

oxidation

▲ **Figure 6** *$Mg^{2+}(aq)|Mg$ has a very negative $E^\ominus$ value and a greater tendency to lose electrons*

reduction

$Cl_2(g) + 2e^- \rightleftharpoons 2Cl^-(aq)\ E = +1.36$ V

oxidation

▲ **Figure 7** *$Cl_2(g)$, $Cl^-(aq)$ has a very positive $E^\ominus$ value and a greater tendency to gain electrons*

Measuring cell potentials

Cells can easily be assembled using any half-cells. The e.m.f. measured is then a cell potential, E_{cell}.

Figure 8 shows a standard cell made from $Zn^{2+}(aq)|Zn(s)$ and $Cu^{2+}(aq)|Cu(s)$ half-cells. Notice that the $Cu^{2+}(aq)$ and $Zn^{2+}(aq)$ solutions both have standard concentrations of 1 mol dm^{-3}.

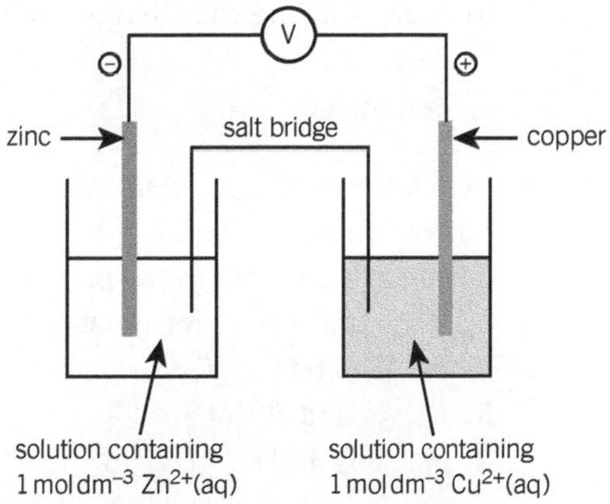

▲ **Figure 8** *A standard cell made from $Zn^{2+}(aq)|Zn(s)$ and $Cu^{2+}(aq)|Cu(s)$ half-cells*

- The copper half-cell has the more positive $E^\ominus$ of +0.34 V, and has a greater tendency to undergo reduction and to gain electrons.
- The zinc half-cell has the more negative $E^\ominus$ of −0.76 V, and a greater tendency to undergo oxidation and to lose electrons.

Electrons flow along the wire from the more negative zinc half-cell to the less negative copper half-cell. The zinc electrode is negative and the copper electrode is positive.

Revision tip

The vertical line, |, indicates the phase boundary between the solid electrode and the aqueous solution.

Revision tip

In an ion/ion half-cell, it may be difficult to dissolve enough solute to get up to concentrations of 1 mol dm^{-3}.

Equal ion concentrations (equimolar), with concentrations less than 1 mol dm^{-3}, are often used which give the same e.m.f.

Revision tip
When combining two half-equations from electrode potentials, the oxidation half equation (with the more negative $E^\ominus$ value) is reversed. In the copper–zinc cell, the oxidation half-equation for Zn^{2+}/Zn is reversed.

Synoptic link
For details of combining two half equations into an overall equation, see Topic 23.1, Redox reactions.

Revision tip
To calculate $E^\ominus_{cell}$, look up the $E^\ominus$ values (see Table 1) and calculate the difference.

Writing an equation for the overall cell reaction

The equilibria for redox systems are written so that the forward reaction shows reduction and the reverse reaction shows oxidation (see Table 1).

The overall cell equation is obtained by combining the reduction and oxidation half-equations.

- The more positive copper half-cell undergoes reduction, gaining electrons and reacting from left to right.
- The more negative zinc half-cell undergoes oxidation, losing electrons and reacting from right to left.

The equation for the more negative redox system undergoes oxidation and is reversed. The two half-equations are then combined and electrons are balanced.

reduction:	$Cu^{2+}(aq) + 2e^- \rightarrow Cu(s)$	positive electrode
oxidation:	$Zn(s) \rightarrow Zn^{2+}(aq) + 2e^-$	negative electrode
overall cell reaction:	$Zn(s) + Cu^{2+}(aq) \rightarrow Zn^{2+}(aq) + Cu(s)$	

Calculating cell potentials from electrode potentials

Standard electrode potentials quantify the tendency of redox systems to gain or lose electrons. A standard cell potential, $E^\ominus_{cell}$, can be calculated directly from standard electrode potentials. The calculation is simply the difference between the $E^\ominus$ values:

$$E^\ominus_{cell} = E^\ominus(\text{positive electrode}) - E^\ominus(\text{negative electrode})$$

For a standard zinc–copper cell,

$$E^\ominus_{cell} = +0.34 - (-0.76) = 1.10\ \text{V}$$

Summary questions

1 a What is meant by 'standard electrode potential'. (Include standard conditions.) *(3 marks)*

b In a cell, what are the charge carriers in the:

i wire *(1 mark)*

ii salt bridge? *(1 mark)*

2 a In a cell, how can you work out the polarity (+ or –) of each half-cell from $E^\ominus$ values? *(1 mark)*

b Calculate the standard cell potential for the following cells from the redox systems in Table 1:

i $Fe^{2+}|Fe$ and $Cu^{2+}|Cu$ *(1 mark)*

ii $Ag^{+}|Ag$ and $Al^{3+}|Al$ *(1 mark)*

iii $Mg^{2+}|Mg$ and $Fe^{3+},Fe^{2+}|Pt$. *(1 mark)*

3 For each cell in question 2, write half-equations for the oxidation and reduction reactions and the overall cell reaction. *(9 marks)*

23.5 Predictions from electrode potentials

Specification reference: 5.2.3

Predicting redox reactions from electrode potentials

You can predict the feasibility of redox reactions from standard electrode potentials.

Table 1 shows three redox systems sorted with the most negative $E^\ominus$ value at the top. The forward reaction is reduction.

- The strongest oxidising agent is at the bottom on the left.
- The strongest reducing agent is at the top on the right.

From Table 1, we can predict that a redox reaction *should* take place:

- between an oxidising agent on the left and
- a reducing agent on the right,

provided that the redox system of the oxidising agent has a more positive $E^\ominus$ value than the redox system of the reducing agent.

Predicting reactions

You can predict potential redox reactions by comparing the strengths of oxidising and reducing agents.

Redox system C

In Table 1, redox system **C** has the **most** positive $E^\ominus$ value and has a greater tendency to be reduced than redox systems **A** and **B**.

We can predict that $Ag^+(aq)$ (the oxidising agent on the left of **C**) would oxidise both Cr(s) and Cu(s) (the reducing agents on the right in redox systems **A** and **B**).

oxidation	**A** $Cr^{3+}(aq) + 3e^-$	$\longleftarrow$	Cr(s)	$E^\ominus = -0.77\,V$
oxidation	**B** $Cu^{2+}(aq) + 2e^-$	$\longleftarrow$	Cu(s)	$E^\ominus = +0.34\,V$
reduction	**C** $Ag^+(aq) + e^-$	$\longrightarrow$	Ag(s)	$E^\ominus = +0.80\,V$

You can write overall equations for the two feasible reactions by

- reversing the oxidation half equation (with the more negative $E^\ominus$ value)
- combining the reduction and oxidation half equations and balancing electrons:

$$3Ag^+(aq) + Cr(s) \rightarrow 3Ag(s) + Cr^{3+}(aq)$$

$$2Ag^+ + Cu(s) \rightarrow 2Ag(s) + Cu^{2+}(aq)$$

Redox system B

Redox system **B** has a more positive $E^\ominus$ value than **A** and will have a greater tendency to be reduced.

We can predict that $Cu^{2+}(aq)$ (the oxidising agent on the left of **B**) would oxidise Cr(s) (the reducing agent on the right in redox system **A**).

oxidation	**A** $Cr^{3+}(aq) + 3e^-$	$\longleftarrow$	Cr(s)	$E^\ominus = -0.77\,V$
reduction	**B** $Cu^{2+}(aq) + 2e^-$	$\longrightarrow$	Cu(s)	$E^\ominus = +0.34\,V$
overall	$3Cu^{2+}(aq) + 2Cr(s) \rightarrow 3Cu(s) + 2Cr^{3+}(aq)$			

Synoptic link

You also met the idea of feasibility with ΔG. See Topic 22.5, Free energy.

▼ **Table 1** *Standard electrode potentials: reducing and oxidising agents*

	Redox system	$E^\ominus$/V
	oxidation ⟵ / reducing agent	
A	$Cr^{3+}(aq) + 3e^- \rightleftharpoons Cr(s)$	−0.77
B	$Cu^{2+}(aq) + 2e^- \rightleftharpoons Cu(s)$	+0.34
C	$Ag^+(aq) + e^- \rightleftharpoons Ag(s)$	+0.80
	oxidising agent / ⟶ reduction	

Revision tip

- The most negative system has the greatest tendency to be oxidised and lose electrons.
- The most positive system has the greatest tendency to be reduced and gain electrons.

Revision tip

You could carry out this process by comparing the strengths of reducing agents. The logic is the same.

Revision tip

You can predict redox reactions by listing the redox systems in electrode potential order, with the most negative $E^\ominus$ value at the top:

- An oxidising agent on the left reacts **only** with
- reducing agents on the right that are **above in the list.**

Synoptic link

For details of combining two half-equations into an overall equation, see Topic 23.1, Redox reactions, and Topic 23.4, Electrode potentials.

Redox system A

Redox system **A** has a **less** positive $E^{\ominus}$ value than redox systems **B** and **C**.

We can predict that

- Cr^{3+}(aq) (the oxidising agent on the left of **A**) would **not** react with the reducing agents in **B** and **C**.

Limitations of feasibility predictions

Standard electrode potentials are useful for predicting feasibility, but predicted redox reactions often do not take place in practice.

Synoptic link

You came across a similar idea with predictions from ΔG values in Topic 22.5, Free energy.

Reaction rate

A predicted reaction may have a large activation energy, resulting in a very slow rate. $E^{\ominus}$ values may indicate the feasibility of a reaction but they give no indication of the rate of a reaction.

Concentration

Predictions are based on standard electrode potentials measured using concentrations of 1 mol dm^{-3}. For concentrations that are not 1 mol dm^{-3}, the electrode potential is different from the standard value, $E^{\ominus}$.

Other limitations

- Standard conditions of temperature and pressure also apply to $E^{\ominus}$ values.
- Standard electrode potentials apply to aqueous equilibria. Many reactions are not aqueous.

Summary questions

1 **a** How can you identify the strongest oxidising agent from redox systems and their $E^{\ominus}$ values? *(2 marks)*

b Why do predictions based on $E^{\ominus}$ values sometimes break down in practice? *(3 marks)*

2 You are provided with the following information.

$Mg^{2+}(aq) + 2e^- \rightleftharpoons Mg(s)$ $E^{\ominus} = -2.37\,V$

$Cr^{3+}(aq) + e^- \rightleftharpoons Cr^{2+}(aq)$ $E^{\ominus} = -0.41\,V$

$I_2(aq) + 2e^- \rightleftharpoons 2I^-(aq)$ $E^{\ominus} = +0.54\,V$

a What is the:

i strongest oxidising agent

ii strongest reducing agent? *(2 marks)*

b **i** Predict the species that would react with Cr^{3+}(aq). *(1 mark)*

ii Predict the species that would react with Mg(s). *(1 mark)*

iii Write equations for the overall reactions in **(ii)**. *(2 marks)*

3 You are provided with the following information.

$CrO_4^{2-}(aq) + 4H_2O(l) + 3e^- \rightleftharpoons Cr(OH)_3(s) + 5OH^-(aq)$ $E^{\ominus} = -0.13\,V$

$IO^-(aq) + H_2O(l) + 2e^- \rightleftharpoons I^-(aq) + 2OH^-$ $E^{\ominus} = +0.49\,V$

$Ag^+(aq) + e^- \rightleftharpoons Ag(s)$ $E^{\ominus} = +0.80\,V$

Predict feasible redox reactions based on these $E^{\ominus}$ values and write equations for the reactions. *(3 marks)*

23.6 Storage and fuel cells

Specification reference: 5.2.3

Modern cells

Modern cells can be: primary cells, secondary cells, or fuel cells. Cells convert chemical energy from redox reactions into electrical energy.

Primary cells

Primary cells are used until the chemicals have reacted when the cell 'goes flat'. The cells cannot be recharged and are discarded after use. The simplest primary cell is the zinc–copper cell (see Topic 23.4, Electrode potentials).

Secondary cells

Secondary cells can be recharged. During recharging, the reaction that 'discharges' the cell during use is reversed.

Revision tip

You do not need to learn details of any specific cell, but you do need to apply the important principles of electrode potentials, half-equations, and cell reactions.

Fuel cells

A fuel cell uses the energy from the reaction of a fuel with oxygen to create a voltage. Fuel cells can operate continuously.

- Oxygen is at the positive electrode where reduction takes place.
- The fuel is at the negative electrode where oxidation takes place.

Hydrogen fuel cell

Hydrogen fuel cells can operate with an acid or alkali electrolyte. With an acid electrolyte, the redox systems and overall reaction are shown below.

$$2H^+(aq) + 2e^- \rightleftharpoons H_2(g) \qquad E^\ominus = 0.00\ V$$

$$O_2(g) + 4H^+(aq) + 4e^- \rightleftharpoons 2H_2O(l) \qquad E^\ominus = +1.23\ V$$

$$\text{overall reaction:}\quad 2H_2(g) + O_2(g) \rightarrow 2H_2O(l) \qquad E_{cell} = 1.23\ V$$

Unlike carbon-containing fossil fuels, hydrogen fuel cells produce no carbon dioxide, with water being the only product. There are obvious environmental benefits with the link between carbon dioxide and global warming.

Synoptic link

For details of combining two half-equations into an overall equation, see Topic 23.1, Redox reactions, and Topic 23.4, Electrode potentials.

Summary questions

1 a What is the difference between a primary cell and a secondary cell? *(1 mark)*
  b What is the key feature of a fuel cell? *(1 mark)*

2 A cell has a cell potential of 1.35 V. The redox systems **A** and **B** in the cell are shown below. **A** is the negative half-cell.
  A $2H_2O + 2e^- \rightleftharpoons H_2 + 2OH^-$ $E^\ominus = -0.83\ V$
  B $NiOOH + H_2O + e^- \rightleftharpoons Ni(OH)_2 + OH^-$
  a Write the equation for the overall cell reaction. *(1 mark)*
  b What is the standard electrode potential of redox system **B**? *(1 mark)*

3 An alkaline aluminium–oxygen fuel cell has $E^\ominus_{cell} = 2.75\ V$.
  The overall cell reaction is: $4Al + 6H_2O + 3O_2 \rightarrow 4Al(OH)_3$
  The oxygen electrode is the positive half-cell:
  $O_2 + 2H_2O + 4e^- \rightleftharpoons 4OH^-$ $E^\ominus = +0.40\ V$
  a What is the standard cell potential of the negative half-cell? *(1 mark)*
  b Explain which half-cell undergoes reduction. *(1 mark)*
  c What is the half-equation that takes place at the oxidation half-cell? *(2 marks)*

Chapter 23 Practice questions

▼ **Table 1** *Half-cells*

half-cell	$E^{\ominus}$/ V
Mg^{2+}|Mg	−2.37
Cd^{2+}|Cd	−0.40

1 A standard cell is set up from the half-cells in Table 1.

What is the half-equation at the positive electrode?

A $Cd^{2+}(aq) + 2e^- \rightarrow Cd(s)$

B $Cd(s) \rightarrow Cd^{2+}(aq) + 2e^-$

C $Mg^{2+}(aq) + 2e^- \rightarrow Mg(s)$

D $Mg(s) \rightarrow Mg^{2+}(aq) + 2e^-$ *(1 mark)*

2 The overall reaction and standard cell potential of a cell is shown below.

$$Mn(s) + 2Ag^+(aq) \rightarrow Mn^{2+}(aq) + 2Ag(s) \qquad E^{\ominus}_{cell} = 1.99\,V$$

For: $Ag^+(aq) + e^- \rightleftharpoons Ag(s)$, the standard electrode potential, $E^{\ominus}$, = +0.80 V.

What is the standard electrode potential, $E^{\ominus}$, for: $Mn^{2+}(aq) + 2e^- \rightleftharpoons Mn(s)$?

A −2.79 V

B −1.19 V

C +1.19 V

D +2.79 V *(1 mark)*

3 Which reaction has the greatest change in oxidation number for nitrogen?

A $4NH_3 + 5O_2 \rightarrow 4NO + 6H_2O$

B $2NO_2 + H_2O \rightarrow HNO_3 + HNO_2$

C $2NO + O_2 \rightarrow 2NO_2$

D $N_2 + 3H_2 \rightarrow 2NH_3$ *(1 mark)*

4 **a** Using oxidation numbers, balance the following equations.

i $Cu + NO_3^- + H^+ \rightarrow Cu^{2+} + NO_2 + H_2O$ *(1 mark)*

ii $MnO_4^- + H^+ + Sn^{2+} \rightarrow Mn^{2+} + Sn^{4+} + H_2O$ *(1 mark)*

b $Cr_2O_7^{2-}$ ions react with I^- ions in the presence of acid, H^+ to form Cr^{3+} ions, I_2, and one other product. Write the balanced equation. *(2 marks)*

5 6.84 g of an iron(II) compound is dissolved in dilute sulfuric acid and made up to 250.0 cm^3 of solution. In a titration, 25.00 cm^3 of this solution reacts exactly with 24.50 cm^3 of 0.0200 mol dm^{-3} $KMnO_4$.

The half-equations are: $Fe^{2+}(aq) \rightarrow Fe^{3+}(aq) + e^-$

$$MnO_4^-(aq) + 8H^+(aq) + 5e^- \rightarrow Mn^{2+}(aq) + 4H_2O(l)$$

a Write the overall equation for the reaction in the titration. *(1 mark)*

b Calculate the percentage by mass of iron in the iron(II) compound. *(5 marks)*

▼ **Table 2** *Redox systems*

	Redox system	$E^{\ominus}$/ V
1	$Al^{3+}(aq) + 3e^- \rightleftharpoons Al(s)$	−1.66
2	$Fe^{2+}(aq) + 2e^- \rightleftharpoons Fe(s)$	−0.44
3	$Cr^{3+}(aq) + e^- \rightleftharpoons Cr^{2+}(aq)$	−0.41
4	$Pb^{2+}(aq) + 2e^- \rightleftharpoons Pb(s)$	−0.13
5	$Mn^{3+}(aq) + e^- \rightleftharpoons Mn^{2+}(aq)$	+1.49

6 This question uses redox systems **1–5** in Table 2.

a **i** What is the strongest oxidising agent? *(1 mark)*

ii What is the strongest reducing agent? *(1 mark)*

b A standard cell is set up from redox systems **1** and **2**.

i What is the cell potential? *(1 mark)*

ii Which half-cell contains the negative electrode? *(1 mark)*

iii Write the overall equation for the cell reaction. *(1 mark)*

c **i** Which species reduces Mn^{3+} but does not reduce Fe^{2+}? *(1 mark)*

ii Write overall equations for the reactions in **i.** *(2 marks)*

24.1 d-block elements

Specification reference: 5.3.1

d-block elements

The d-block elements are in the centre of the periodic table, between Group 2 and Group 13 (3).

Electron configurations of d-block atoms

Figure 1 shows the sub-shell energies of the first four shells. Electrons occupy orbitals in order of increasing energy.

The 4s sub-shell has

- a lower energy than the 3d sub-shell
- is filled before the 3d sub-shell.

Across the periodic table from scandium to zinc, electrons are added to 3d orbitals – hence the name d-block elements. Table 1 shows the electron configurations of atoms of the d-block elements.

▼ **Table 1** *The electron configurations of atoms of d-block elements*

Element	Number of electrons	Electron configuration
scandium	21	$1s^22s^22p^63s^23p^6\mathbf{3d^1}4s^2$
titanium	22	$1s^22s^22p^63s^23p^6\mathbf{3d^2}4s^2$
vanadium	23	$1s^22s^22p^63s^23p^6\mathbf{3d^3}4s^2$
chromium	24	$1s^22s^22p^63s^23p^6\mathbf{3d^54s^1}$
manganese	25	$1s^22s^22p^63s^23p^6\mathbf{3d^5}4s^2$
iron	26	$1s^22s^22p^63s^23p^6\mathbf{3d^6}4s^2$
cobalt	27	$1s^22s^22p^63s^23p^6\mathbf{3d^7}4s^2$
nickel	28	$1s^22s^22p^63s^23p^6\mathbf{3d^8}4s^2$
copper	29	$1s^22s^22p^63s^23p^6\mathbf{3d^{10}4s^1}$
zinc	30	$1s^22s^22p^63s^23p^6\mathbf{3d^{10}}4s^2$

The special case of chromium and copper

The electron configurations of chromium and copper do not follow the trend for the other elements (See Table 1).

- Cr atom: Expected = $[Ar]3d^44s^2$
  Actual = $[Ar]3d^54s^1$ **half-filled** d sub-shell → extra stability.
- Cu atom: Expected = $[Ar]3d^94s^2$
  Actual = $[Ar]3d^{10}4s^1$ **full** d sub-shell → extra stability.

Electron configuration of d-block ions

The d-block elements, scandium to zinc, form positive ions from their atoms. The 4s electrons are lost **before** the 3d electrons.

- When forming an **atom**, the 4s orbital **fills before** the 3d orbitals.
- When forming an **ion**, the 4s orbital **empties before** the 3d orbitals.

e.g.	Co atom	$1s^22s^22p^63s^23p^63d^74s^2$	4s orbitals filled before 3d orbitals
	Co^{2+} ion	$1s^22s^22p^63s^23p^63d^7$	Two 4s electrons removed.
	Co^{3+} ion	$1s^22s^22p^63s^23p^63d^6$	Two 4s and one 3d electron removed

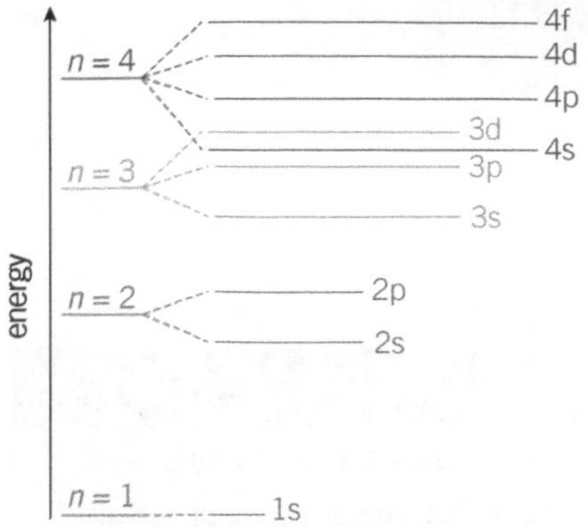

▲ **Figure 1** *Energy level diagram showing the overlap of the 3d and 4s sub-shells*

Synoptic link

See Topic 5.1, Electron structure, for details of filling orbitals.

Revision tip

From Sc to Zn, the d sub-shell is being filled.

The electron configurations are often shortened based on the electron configuration of the previous noble gas,

e.g. Sc is often written as $[Ar]3d^14s^2$

Revision tip

It is believed that a half-full d sub-shell (d^5) and a full d sub-shell (d^{10}) give additional stability to atoms.

Take care with the electron configurations of Cr and Cu.

Revision tip

When forming ions of d-block elements, you remove the 4s electrons before removing any of the 3d electrons.

Notice that a 3d electron is lost for Co^{3+} but only after the 4s electrons have been lost.

Transition elements

Transition elements are d-block elements that form an ion with an incomplete d sub-shell.

The d-block elements scandium and zinc are **not** classified as transition elements because they do not form any ions with a partially filled d-orbital.

Scandium only forms the ion Sc^{3+}:

- Sc atom: $1s^2 2s^2 2p^6 3s^2 3p^6 4s^2 3d^1$
- Sc^{3+} ion: $1s^2 2s^2 2p^6 3s^2 3p^6$ *empty d sub-shell*

Zinc only forms the ion Zn^{2+}:

- Zn atom: $1s^2 2s^2 2p^6 3s^2 3p^6 4s^2 3d^{10}$
- Zn^{2+} ion: $1s^2 2s^2 2p^6 3s^2 3p^6 3d^{10}$ *full d sub-shell*

Properties of transition elements and compounds

The transition elements have characteristic properties that are different from other metals.

- They form compounds with different oxidation states.
- They form coloured compounds.
- The elements and their compounds can act as catalysts.

Variable oxidation state and coloured compounds

Transition metal compounds and ions are often coloured.

The colour is often characteristic of the transition metal ion and its oxidation state. The colour of a solution can vary with a change in oxidation state, ligand, or coordination number.

e.g. Iron(II) Fe^{2+} $1s^2 2s^2 2p^6 3s^2 3p^6 3d^6$ – pale green

Iron(III) Fe^{3+} $1s^2 2s^2 2p^6 3s^2 3p^6 3d^5$ – yellow

Catalytic behaviour

A **catalyst** increases the rate of a chemical reaction by providing an alternative reaction pathway with a lower activation energy.

Transition metals and their compounds are important catalysts for many industrial processes and in the laboratory.

Transition metal catalysts in industry

- Iron in the Haber process for the manufacture of ammonia:

  $N_2(g) + 3H_2(g) \rightleftharpoons 2NH_3(g)$
- Vanadium pentoxide, V_2O_5, in the production of sulfur trioxide for the manufacture of sulfuric acid:

  $2SO_2(g) + O_2(g) \rightleftharpoons 2SO_3(g)$

Transition metal catalysts in the laboratory

- Manganese(IV) oxide, MnO_2, in the decomposition of hydrogen peroxide to form oxygen:

  $2H_2O_2(aq) \rightarrow 2H_2O(l) + O_2(g)$
- $Cu^{2+}(aq)$ ions in the reaction of zinc metal with acids:

  $Zn(s) + H_2SO_4(aq) \rightarrow ZnSO_4(aq) + H_2(g)$

Synoptic link

See Topic 4.3, Redox, and Topic 23.1, Redox reactions, for details of oxidation numbers.

Synoptic link

For more about coordination number and ligands, see Topic 24.2, The formation and shapes of complex ions.

Synoptic link

Catalysis was discussed in detail in Topic 10.2, Catalysts.

Summary questions

1 State three characteristic properties of the transition elements, different from other metals. *(3 marks)*

2 a Write the electron configurations for
   i A Ni atom
   ii a Mn^{3+} ion. *(2 marks)*
  b What is the oxidation state of:
   i Mn in Mn_2O_3
   ii Mn in K_2MnO_4
   iii Fe in FeO_4^{2-}
   iv V in VO_2^+ *(4 marks)*

3 Explain why scandium and zinc are classified as d-block metals but not transition metals. *(4 marks)*

24.2 The formation and shapes of complex ions

Specification reference: 5.3.1

Ligands and complex ions

Ligands

A **ligand** is a molecule or ion that donates a pair of electrons to a central metal ion to form a coordinate bond or dative covalent bond.

A ligand must have a lone pair of electrons on an electronegative atom such as N, O, F, and Cl.

Monodentate ligands

A **monodentate ligand** donates *one* pair of electrons to a central metal ion. See Table 1 for common examples.

▼ **Table 1** *Common monodentate ligands*

Monodentate ligands	
Name	Formula
water	H_2O:
ammonia	:NH_3
chloride	:Cl^-
cyanide	: CN^-
hydroxide	: OH^-

Bidentate ligands

A **bidentate ligand** donates *two* pairs of electrons to a central metal ion. Bidentate ligands contain two electronegative atoms with lone pairs. See Figure 1 for common examples.

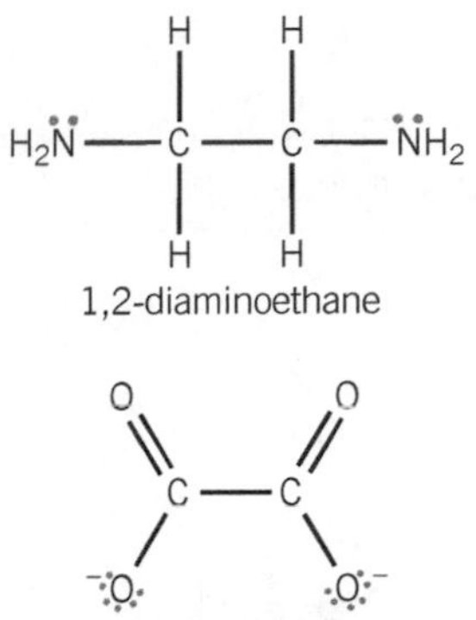

▲ **Figure 1** *Bidentate ligands: 1,2-diaminoethane, $NH_2CH_2CH_2NH_2$, ('en') and ethanedioate (oxalate), $^-OOCCOO^-$*

Complex ions

A **complex ion** is formed when ligands bond to a central metal ion.

The **coordination number** is the number of coordinate bonds attached to the central metal ion.

Complex ions with monodentate ligands

$[Cr(H_2O)_6]^{3+}$ (Figure 2) is a complex ion containing monodentate ligands.

- There are **six** H_2O ligands, each forming **one** coordinate bond.
- The coordination number is **6** and there are **six** coordinate bonds to the central Cr^{3+} ion.
- The metal ion and ligands are shown inside square brackets, [ ].
- The overall charge is 3+ shown outside the square brackets.

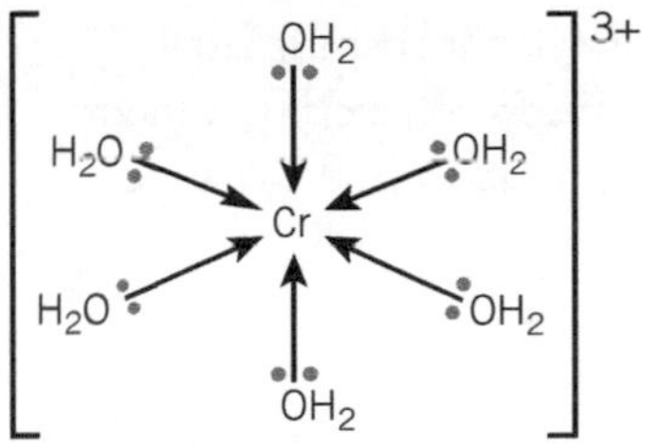

▲ **Figure 2** *The six coordinate bonds in the complex ion $[Cr(H_2O)_6]^{3+}$*

Complex ions with bidentate ligands

$[Co(NH_2CH_2CH_2NH_2)_3]^{3+}$ (Figure 3) is a complex ion containing bidentate ligands.

- There are **three** $NH_2CH_2CH_2NH_2$ ligands, each forming **two** coordinate bonds.
- There are **six** coordinate bonds and the coordination number is **6**. (3 × 2 = 6)
- The overall charge is 3+.

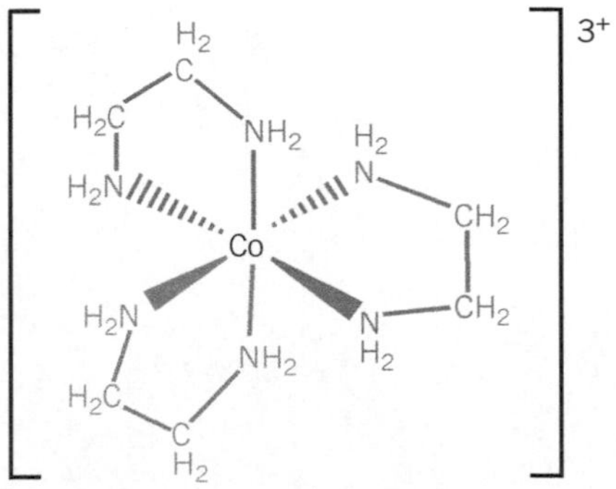

▲ **Figure 3** *The complex ion $[Co(NH_2CH_2CH_2NH_2)_3]^{3+}$ – three bidentate ligands and six coordinate bonds*

Shapes of complex ions

The shape of a complex ion depends upon its coordination number. The commonest coordination numbers are six and four giving rise to six-fold and four-fold coordination.

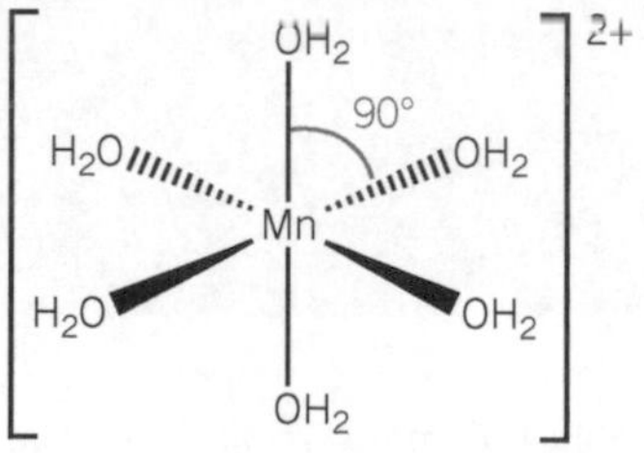

▲ **Figure 4** *$[Mn(H_2O)_6]^{2+}$ is 6-coordinate with an octahedral shape and a bond angle of 90°*

Synoptic link

When drawing shapes, it is essential that you show 3D diagrams including wedges. You should also be able to explain the shapes in terms of electron pair repulsion.

For details, see Topic 6.1, Shapes of molecules and ions.

Six-fold coordination

Many complex ions have a coordination number of six, giving an octahedral shape.

An octahedral shape (Figure 4) has:

- the ligands arranged at the corners of an octahedron
- bond angles of 90° around the central metal ion.

Four-fold coordination

Complex ions with a coordination number of four have two common shapes: tetrahedral and square planar.

Tetrahedral complexes

A tetrahedral shape (Figure 5) has:

- the ligands arranged at the corners of a tetrahedron
- bond angles of 109.5° around the central metal ion.

$[CoCl_4]^{2-}$ $[CuCl_4]^{2-}$

▲ **Figure 5** *$[CoCl_4]^{2-}$ and $[CuCl_4]^{2-}$ have a tetrahedral shape with a bond angle of 109.5°*

Square planar complexes

A square planar shape occurs in complexes of transition metals with eight d-electrons in the highest energy d-sub-shell, e.g. platinum(II) and palladium(II).

A square planar shape (Figure 6) has:

- the ligands arranged at the corners of a square
- bond angles of 90° around the central metal ion.

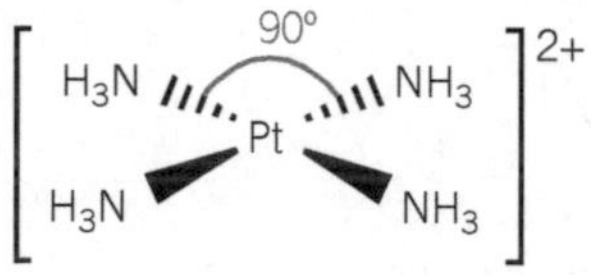

▲ **Figure 6** *$[Pt(NH_3)_4]^{2+}$ has a square planar shape with a bond angle of 90°*

Revision tip

A square planar shape is similar to the octahedral shape, but without the ligands above and below the plane.

Summary questions

1 **a** What is meant by the terms:
  i ligand? **ii** bidentate ligand? **iii** coordination number? *(3 marks)*
b What is the coordination number and bond angle in $[Ni(NH_2CH_2CH_2NH_2)_3]^{2+}$? *(2 marks)*

2 **a** State the shape, bond angle, and coordination number in
  i $[Cr(NH_3)_6]^{3+}$ **ii** $[Cu(Cl)_4]^{2-}$ *(6 marks)*
b What is the oxidation number of the metal in the following?
  i $[Pt(NH_3)_3Cl]^+$ **ii** $[Cr(H_2O)_4Cl_2]^+$ *(2 marks)*
c What is the formula of the complex ion containing the following?
  i Cu^{2+} and 6 F^- ligands *(1 mark)*
  ii Fe^{3+} and 6 CN^- ions *(1 mark)*
  iii Co^{3+}, 4 NH_3 and 2 Cl^- ligands. *(1 mark)*
  iv Ni^{2+} and 3 $H_2NCH_2CH_2NH_2$ ligands *(1 mark)*

3 **a** What is the formula of the 6-coordinate complex ion between Fe^{3+} and $(COO^-)_2$ ligands? [$(COO^-)_2$ is bidentate.] *(1 mark)*
b Ni^{2+} forms a neutral square planar complex with NH_3 and Cl^- ligands. State its formula. *(1 mark)*

24.3 Stereoisomerism in complex ions

Specification reference: 5.3.1

Stereoisomers

Stereoisomers have the same structural formula but a different arrangement of the atoms in space.

Complex ions can display two types of stereoisomerism:

- *cis–trans* isomerism
- optical isomerism.

In complex ions, the type of stereoisomerism depends on the number and type of ligands, and the shape.

Cis–trans isomerism

Cis–trans isomerism occurs in some square planar and octahedral complex ions.

Square planar complexes

Cis and *trans* isomers exist in square planar complexes with:

- **two** molecules or ions of one monodentate ligand, X
- **two** molecules or ions of another monodentate ligand, Y.

The general formula is $[MX_2Y_2]^n$ (n = charge).

Figure 1 shows the *cis* and *trans* isomers of platin, $[Pt(NH_3)_2Cl_2]$.

In the *cis*-isomer:

- the identical ligands are on the same side of the complex, next to each other
- the coordinate bonds between identical ligands are 90° apart.

In the *trans*-isomer:

- the identical ligands are on opposite sides of the complex
- the coordinate bonds between identical ligands are 180° apart.

Cis-platin and cancer treatment

Cis-platin (Figure 1) is used in chemotherapy as an anti-cancer drug.

Cis-platin works by forming a platinum complex, which binds to DNA and prevents cell division. Unfortunately, *cis*-platin and other platinum-based drugs cause unpleasant side effects.

Octahedral complexes

Monodentate ligands

Cis and *trans* isomers exist in octahedral complexes with:

- **four** molecules or ions of one monodentate ligand, X
- **two** molecules or ions of another monodentate ligand, Y.

The general formula is $[MX_4Y_2]^n$ (n = charge).

Figure 2 shows the *cis* and *trans* isomers of the $[Co(NH_3)_4Cl_2]^+$ complex ion.

In the *cis*-isomer:

- the two Cl^- ligands are next to each other
- the coordinate bonds between the Cl^- ions are 90° apart.

In the *trans*-isomer:

- the two Cl^- ligands are opposite each other
- the coordinate bonds between the Cl^- ions are 180° apart.

Synoptic link

In Topic 13.2, Stereoisomerism, you met *cis–trans* and *E/Z* stereoisomerism in alkenes. You will meet optical isomerism in organic chemistry in Topic 27.2, Amino acids, amides, and chirality.

Synoptic link

For details of square planar and octahedral complex ions, see Topic 24.2, The formation and shapes of complex ions.

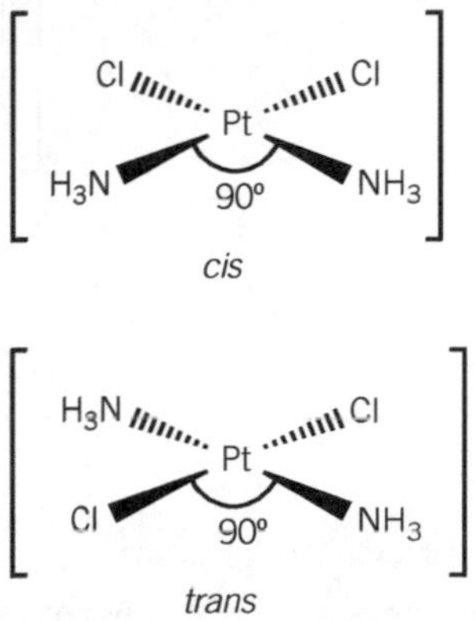

▲ **Figure 1** *The cis and trans isomers of platin, $[Pt(NH_3)_2Cl_2]$*

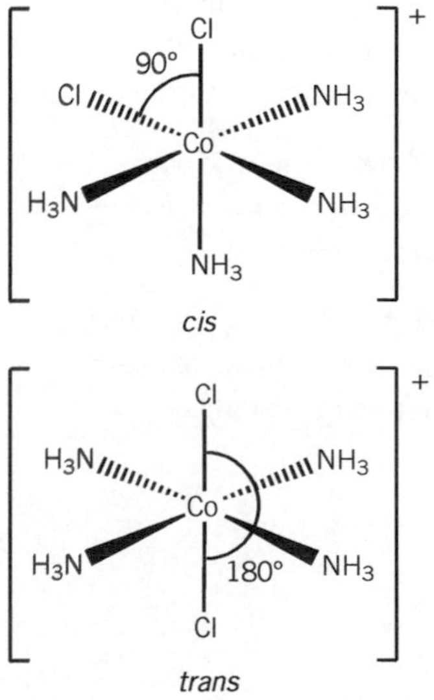

▲ **Figure 2** *The cis and trans isomers of $[Co(NH_3)_4Cl_2]^+$*

Optical isomerism

Optical isomerism only occurs in octahedral complexes containing two or three bidentate ligands.

Revision tip

Optical isomers are mirror images that cannot be superimposed upon each other, rather like a left hand and a right hand.

Revision tip

Trans-isomers of octahedral complexes cannot form optical isomers, as mirror images are the same and can be superimposed.

Optical isomerism with two bidentate ligands

Some six-coordinate complex ions containing monodentate and bidentate ligands can show both *cis–trans* **and** optical isomerism.

Cis isomers of octahedral complexes have optical isomers if they have:

- **two** molecules or ions of a bidentate ligand
- **two** molecules or ions of a monodentate ligand.

The *cis* and *trans* isomers of the $Co(NH_2CH_2CH_2NH_2)_2Cl_2]^+$ complex ion are shown in Figure 3.

The *cis* isomer also has optical isomers, also shown in Figure 3.

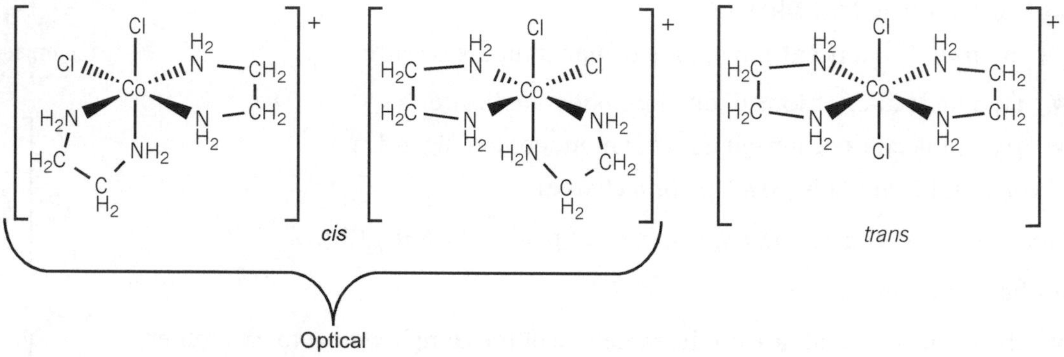

▲ **Figure 3** *The cis-trans and optical isomers of* $[Co(NH_2CH_2CH_2NH_2)_2Cl_2]^+$

Revision tip

Diagrams of optical isomers of complex ions are complicated. You need to take great care:

- First draw a 3D structure, making use of wedges, of one of the isomers. Make sure all bonds are between the metal ion and the ligand atoms with the lone pairs. In this example, the N atoms have the lone pairs.
- Now draw the second optical isomer as the mirror image of the first isomer.
- Finally check that you haven't missed anything out or made any errors in bonding.

Optical isomerism with three bidentate ligands

Figure 4 shows the optical isomers of the $[Ni(NH_2CH_2CH_2NH_2)_3]^{2+}$ complex ion, which contains three molecules of the bidentate ligand $NH_2CH_2CH_2NH_2$.

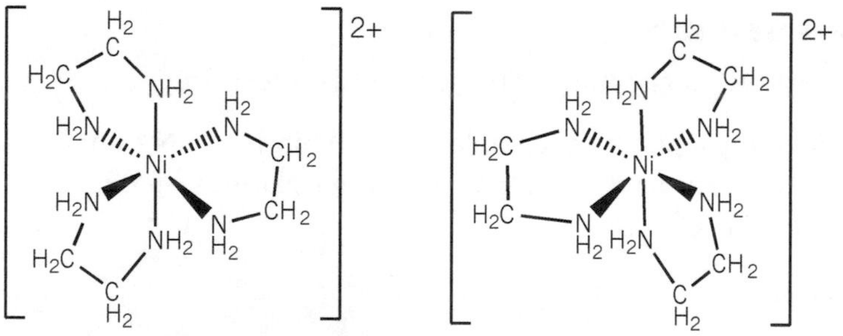

▲ **Figure 4** *The two optical isomers of* $[Ni(NH_2CH_2CH_2NH_2)_3]^{2+}$

Summary questions

1 Describe the role of *cis*-platin in medicine. *(2 marks)*

2 Draw 3D structures of the following.

a The *cis-trans* isomers of $[Mn(NH_3)_4(H_2O)_2]^{2+}$ *(2 marks)*

b The optical isomers of $[Co(NH_2CH_2CH_2NH_2)_3]^{2+}$ *(2 marks)*

3 Mn forms a complex ion with two $(COO^-)_2$ ligands and two H_2O ligands.

a Write the formula of this complex ion. *(1 mark)*

b Draw 3D diagrams of the types of isomerism in this complex ion. *(3 marks)*

24.4 Ligand substitution and precipitation

Specification reference: 5.3.1

Ligand substitution reactions

Ligand substitution is a reversible reaction of a complex ion in which one ligand is replaced by another ligand.

Ligand substitution reactions of $[Cu(H_2O)_6]^{2+}$

Aqueous ammonia, $NH_3(aq)$

$[Cu(H_2O)_6]^{2+}$ reacts with an excess of aqueous ammonia by ligand substitution.

- **Four** H_2O molecules are replaced by **four** NH_3 ligands.
- The colour of the solution changes from pale blue to dark blue.

$$[Cu(H_2O)_6]^{2+} + 4NH_3(aq) \rightleftharpoons [Cu(NH_3)_4(H_2O)_2]^{2+} + 4H_2O(l)$$

pale-blue solution — **dark-blue** solution
octahedral — **octahedral**

> **Revision tip**
> The reaction with aqueous ammonia first forms a precipitate.
>
> See later in this topic in 'Precipitation reactions'.

Aqueous chloride ions, $Cl^-(aq)$

$[Cu(H_2O)_6]^{2+}$ reacts by ligand substitution with an excess of chloride ions. HCl(aq) is used to supply a high concentration of $Cl^-(aq)$ ions.

- **Six** H_2O molecules are replaced by **four** Cl^- ligands.
- The colour of the solution changes from pale-blue to yellow.

$$[Cu(H_2O)_6]^{2+} + 4Cl^-(aq) \rightleftharpoons [CuCl_4]^{2-} + 6H_2O(l)$$

pale-blue solution — **yellow** solution
octahedral — **tetrahedral**

> **Revision tip**
> If the chloride concentration is not high enough, an equilibrium sets up between the two sides of the reaction and the colour seen is green (as a mixture of blue and yellow).

Ligand substitution reactions of $[Cr(H_2O)_6]^{3+}$

Aqueous ammonia, $NH_3(aq)$

$[Cr(H_2O)_6]^{3+}$ reacts by ligand substitution with an excess of aqueous ammonia.

- **Six** H_2O molecules are replaced by **six** NH_3 ligands to form $[Cr(NH_3)_6]^{3+}$.
- The colour of the solution changes from violet to purple.

$$[Cr(H_2O)_6]^{3+} + 6NH_3(aq) \rightleftharpoons [Cr(NH_3)_6]^{3+} + 6H_2O(l)$$

violet solution — **purple** solution
octahedral — **octahedral**

> **Revision tip**
> You are expected to know the violet to purple colour change for this reaction.
>
> If chromium(III) sulfate or chromium(III) chloride is dissolved in water, the initial colour may be dark green from complexing with some sulfate or chloride ions.

Ligand substitution in haemoglobin

Haemoglobin in red blood cells contains four protein chains.

- Each protein chain has a planar haem molecule within its structure.
- The Fe^{2+} metal ion in haem bonds to a protein chain and water.
- The water can exchange readily with oxygen gas, O_2 (Figure 1).

As blood passes through the lungs, haemoglobin bonds to oxygen, forming the deep red complex **oxyhaemoglobin**. The oxygen is released to body cells as and when required, exchanging with water. A simplified equation is shown below (Hb = haemoglobin; HbO_2 = oxyhaemoglobin):

$$Hb + O_2 \rightleftharpoons HbO_2$$

Haemoglobin can also bond to carbon dioxide, which is carried back to the lungs to be exhaled, releasing the carbon dioxide.

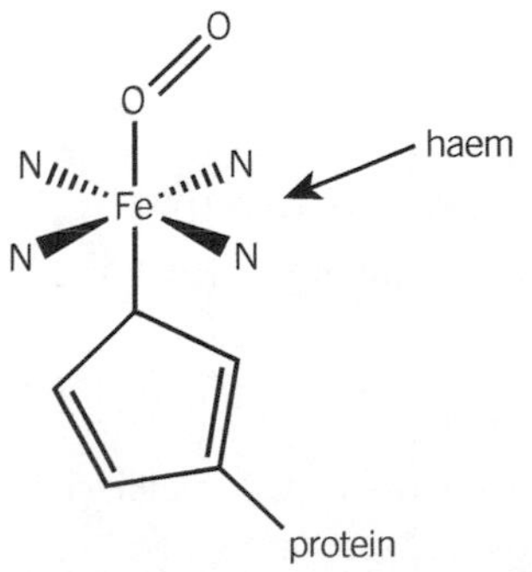

▲ **Figure 1** *Oxygen binds to the Fe^{2+} ion in haem by the formation of a coordinate bond. For clarity, the 2+ charge and other bonds from N atoms have been omitted*

> **Revision tip**
> Every year, deaths are caused by carbon monoxide poisoning of the blood in this way, often caused by inadequate ventilation in heating devices.

Carbon monoxide, CO

Carbon monoxide can also bond to the Fe^{2+} ion in haemoglobin, forming a deep pink complex called carboxyhaemoglobin.

A ligand substitution reaction takes place where the oxygen in oxyhaemoglobin is replaced by carbon monoxide. The bond is so strong that this process is irreversible. If the carbon monoxide concentration is high, the blood is starved of oxygen, leading to death.

Precipitation reactions

A **precipitation reaction** is a reaction between ions in two aqueous solutions to form an insoluble ionic solid (the precipitate).

Solutions of NaOH(aq) and NH_3(aq) both contains hydroxide ions, OH^-(aq).

- Aqueous transition metal ions react with NaOH(aq) and NH_3(aq) to form precipitates of the metal hydroxide.
- Some of these precipitates dissolve in an excess of NaOH(aq) and NH_3(aq) to form complex ions in solution.

Precipitation with OH^- ions (from NaOH(aq) and NH_3(aq))

With NaOH(aq) and NH_3(aq), Cu^{2+}(aq), Fe^{2+}(aq), Fe^{3+}(aq), Mn^{2+}(aq) and Cr^{3+}(aq) ions all form precipitates of the metal hydroxide.

> **Revision tip**
> You need to learn all these precipitation reactions and colours.
>
> You must also learn which metal hydroxides dissolve in excess NaOH(aq) and NH_3(aq), the colour changes, and the equations.
>
> This involves some hard learning of chemistry!

Cu^{2+}

$Cu^{2+}(aq) + 2OH^-(aq) \rightarrow Cu(OH)_2(s)$

pale-blue solution → **blue** precipitate

Fe^{2+}

$Fe^{2+}(aq) + 2OH^-(aq) \rightarrow Fe(OH)_2(s)$

pale-green solution → **green** precipitate

The green precipitate of $Fe(OH)_2$ turns brown at its surface on standing because iron(II) is oxidised to iron(III) by contact with the air:

$Fe(OH)_2(s) \rightarrow Fe(OH)_3(s)$

Fe^{3+}

$Fe^{3+}(aq) + 3OH^-(aq) \rightarrow Fe(OH)_3(s)$

pale-yellow solution → **orange-brown** precipitate

Mn^{2+}

$Mn^{2+}(aq) + 2OH^-(aq) \rightarrow Mn(OH)_2(s)$

pale-pink solution → **light-brown** precipitate

Cr^{3+}

$Cr^{3+}(aq) + 3OH^-(aq) \rightarrow Cr(OH)_3(s)$

violet solution → **green** precipitate

Complex ion formation with excess NaOH(aq) and NH_3(aq)

Excess NaOH(aq)

> **Revision tip**
> With excess NaOH, only $Cr(OH)_3$ reacts further forming $[Cr(OH)_6]^{3-}$.

- The hydroxides of Cu^{2+}, Fe^{2+}, Fe^{3+} and Mn^{2+} do **not** dissolve in excess NaOH(aq).
- The hydroxide of Cr^{3+}(aq) **does** dissolve in excess NaOH(aq).

The reactions of Cr^{3+}(aq) with NaOH(aq) are summarised below.

$$Cr^{3+}(aq) \xrightarrow{NaOH(aq)} Cr(OH)_3(s) \xrightarrow{\text{excess } NaOH(aq)} [Cr(OH)_6]^{3-}$$

violet solution → **green** precipitate → **dark green** solution

> **Revision tip**
> Depending on the Cr^{3+} salt dissolved, Cr^{3+}(aq) may appear as a green solution.

Excess $NH_3(aq)$

- The hydroxides of Fe^{2+}, Fe^{3+} and Mn^{2+} do **not** dissolve in excess $NH_3(aq)$.
- The hydroxides of Cu^{2+} and Cr^{2+} **do** dissolve in excess $NH_3(aq)$.

The reactions of $Cu^{2+}(aq)$ and $Cr^{3+}(aq)$ with $NH_3(aq)$ are summarised below.

$$\underset{\textbf{violet}\text{ solution}}{Cr^{3+}(aq)} \xrightarrow{NH_3(aq)} \underset{\textbf{green}\text{ precipitate}}{Cr(OH)_3(s)} \xrightarrow{\text{excess } NH_3(aq)} \underset{\textbf{purple}\text{ solution}}{[Cr(NH_3)_6]^{3+}}$$

$$\underset{\textbf{pale-blue}\text{ solution}}{Cu^{2+}(aq)} \xrightarrow{NH_3(aq)} \underset{\textbf{blue}\text{ precipitate}}{Cu(OH)_2(s)} \xrightarrow{\text{excess } NH_3(aq)} \underset{\textbf{dark-blue}\text{ solution}}{[Cu(NH_3)_4(H_2O)_2]^{2+}}$$

Revision tip

With excess NH_3:

- $Cr(OH)_3$ reacts further, forming $[Cr(NH_3)_6]^{3+}$.
- $Cu(OH)_2$ reacts further, forming $[Cu(NH_3)_4(H_2O)_2]^{2+}$

(Further details are described under 'Ligand substitution' earlier in this topic.)

Summary questions

1 **a** What is meant by the term ligand substitution? *(1 mark)*

b What is the colour of?

i $[Cr(H_2O)_6]^{3+}$ *(1 mark)*

ii $[Cr(NH_3)_6]^{3+}$ *(1 mark)*

2 For the following reactions, write an ionic equation and state the colour of the precipitate formed.

a $Mn^{2+}(aq)$ and $NaOH(aq)$ *(2 marks)*

b $Cu^{2+}(aq)$ and $NaOH(aq)$ *(2 marks)*

c $Fe^{3+}(aq)$ and $NH_3(aq)$ *(2 marks)*

3 A solution **X** containing complex ion **A** is made by dissolving copper(II) sulfate in water.
Addition of concentrated HCl to solution **X** forms a solution containing complex ion **B**.
Addition of $NH_3(aq)$ to solution **X** forms a precipitate **C**.
With excess $NH_3(aq)$, precipitate **C** dissolves forming a solution containing the complex ion **D**.

a What are the colours and formulae of the complex ions **A**, **B**, and **D**, and precipitate **C**? *(8 marks)*

b State the type of reaction that forms **B** and **D** from **A**. *(1 mark)*

c Write equations for:

i the formation of **B** from **A** *(1 mark)*

ii the formation of **D** from **A**. *(1 mark)*

24.5 Redox and qualitative analysis

Specification reference: 5.3.1

Redox reactions of transition metal ions

Transition metals can form ions with different oxidation states. We can carry out reactions to move between these oxidation states by adding suitable oxidising and reducing agents.

Redox reactions of Fe^{2+} and Fe^{3+}

Oxidation of Fe^{2+} to Fe^{3+}

$Fe^{2+}(aq)$ ions are oxidised to $Fe^{3+}(aq)$ ions by acidified manganate(VII) ions, $MnO_4^-(aq)$:

$$MnO_4^-(aq) + 8H^+(aq) + 5Fe^{2+}(aq) \rightarrow Mn^{2+}(aq) + 5Fe^{3+}(aq) + 4H_2O(l)$$

purple **colourless**

The pale-green colour of $Fe^{2+}(aq)$ is obscured by purple $MnO_4^-(aq)$ ions. The $Fe^{3+}(aq)$ ions are in such low concentration that its yellow colour cannot usually be seen.

- Fe is oxidised from +2 in Fe^{2+} to +3 in Fe^{3+}
- Mn is reduced from +7 in MnO_4^- to +2 in Mn^{2+}

Reduction of Fe^{3+} to Fe^{2+}

$Fe^{3+}(aq)$ ions are reduced to $Fe^{2+}(aq)$ ions by iodide ions, $I^-(aq)$:

$$2Fe^{3+}(aq) + 2I^-(aq) \rightarrow 2Fe^{2+}(aq) + I_2(aq)$$

yellow-orange **pale-green** **brown**

The colour change from yellow-orange for $Fe^{3+}(aq)$ to pale-green for Fe^{2+} is obscured by the formation of iodine, $I_2(aq)$, which has a brown colour.

- Fe is reduced from +3 in Fe^{3+} to +2 in Fe^{2+}
- I is oxidised from −1 in I^- to 0 in I_2

Redox reactions of $Cr_2O_7^{2-}$ and Cr^{3+}

Reduction of $Cr_2O_7^{2-}$ to Cr^{3+}

Acidified dichromate(VI) ions, $Cr_2O_7^{2-}(aq)$ are reduced to chromium(III) ions, $Cr^{3+}(aq)$, by zinc metal:

$$Cr_2O_7^{2-}(aq) + 14H^+(aq) + 3Zn(s) \rightarrow 2Cr^{3+}(aq) + 7H_2O(l) + 3Zn^{2+}(aq)$$

orange **green**

- Cr is reduced from +6 in $Cr_2O_7^{2-}$ to +3 in Cr^{3+}
- Zn is oxidised from 0 in Zn to +2 in Zn^{2+}

Oxidation of Cr^{3+} to CrO_4^{2-}

Chromium(III) ions, $Cr^{3+}(aq)$, are oxidised to chromate(IV) ions, $CrO_4^{2-}(aq)$, by hot alkaline hydrogen peroxide, $H_2O_2(aq)$:

$$2Cr^{3+}(aq) + 3H_2O_2(aq) + 10OH^-(aq) \rightarrow 2CrO_4^{2-}(aq) + 8H_2O(l)$$

yellow solution

- Chromium is oxidised from +3 in Cr^{3+} to +6 in CrO_4^{2-}
- Oxygen is reduced from −1 in H_2O_2 to −2 in CrO_4^{2-}

Synoptic link

For more details of the redox reaction between acidified manganate(VII) ions, $MnO_4^-(aq)$, and iron(II) ions, $Fe^{2+}(aq)$ see Topic 23.2, Manganate(VII) redox titrations.

Revision tip

The colour of $Fe^{3+}(aq)$ depends very much on concentration and varies from pale yellow (low concentrations) through yellow-orange to orange-brown for high concentrations.

Very dilute solutions of both Fe^{2+} and Fe^{3+} may appear colourless.

Revision tip

The colour of $Cr^{3+}(aq)$ seen can vary. The $[Cr(H_2O)_6]^{3+}$ ion is violet, but it often appears a green colour especially in the presence of H_2SO_4.

In this reaction, $H_2SO_4(aq)$ is usually used, as the acid and a green colour is seen.

Redox reactions of Cu^{2+} and Cu^+

Reduction of Cu^{2+} to Cu^+

Copper(II) ions, $Cu^{2+}(aq)$, are reduced to solid copper(I) iodide, CuI(s), by iodide ions, $I^-(aq)$.

$2Cu^{2+}(aq)$	+	$4I^-(aq)$	→	$2CuI(s)$	+	$I_2(aq)$
pale-blue				**white**		**brown**
solution				precipitate		solution

- Cu is reduced from +2 in Cu^{2+} to +1 in CuI
- I is oxidised from −1 in I^- to 0 in I_2

Disproportionation of Cu^+ ions

Solid copper(I) oxide, Cu_2O, reacts with hot dilute sulfuric acid to form a brown precipitate of copper metal and a blue solution of copper(II) sulfate.

This is a disproportionation reaction as copper(I) ions, Cu^+, have been simultaneously reduced to copper, Cu, and oxidised to copper(II) ions, Cu^{2+}.

$Cu_2O(s)$	+	$H_2SO_4(aq)$	→	$Cu(s)$	+	$CuSO_4(aq)$	+	$H_2O(l)$
red-brown				**brown**		**blue**		
solid				solid		solution		

- Cu is reduced from +1 in Cu_2O to 0 in Cu
- Cu is oxidised from +1 in Cu_2O to +2 in $CuSO_4$

Qualitative analysis of ions

Qualitative analysis of ions can be carried out on a test-tube scale as a convenient way of identifying anions and cations in an unknown compound.

You have covered tests for the following anions and cations:

- anions: CO_3^{2-}, Cl^-, Br^-, I^-, SO_4^{2-}
- cations: NH_4^+, Cu^{2+}, Fe^{2+}, Fe^{3+}, Mn^{2+}, Cr^{3+}.

Summary questions

1 a State the colours of the following aqueous ions.
a $Fe^{2+}(aq)$ **b** $Fe^{3+}(aq)$ **c** $Cr_2O_7^{2-}(aq)$ **d** $CrO_4^{2-}(aq)$ *(4 marks)*

2 a State an oxidising agent and the oxidation number changes for the following conversions.
i $Fe^{2+}(aq) \rightarrow Fe^{3+}(aq)$ *(3 marks)*
ii $Cr^{3+}(aq) \rightarrow CrO_4^{2-}(aq)$ *(3 marks)*

b State a reducing agent and the oxidation number changes for the following conversions.
i $Fe^{3+}(aq) \rightarrow Fe^{2+}(aq)$ *(3 marks)*
ii $Cr_2O_7^{2-}(aq) \rightarrow Cr^{3+}(aq)$ *(3 marks)*
iii $Cu^{2+} \rightarrow Cu^+$ *(3 marks)*

3 a Describe the disproportionation of copper(I) ions, including an equation and colours. *(4 marks)*

b Describe simple chemical tests that give different observations for the following. Include products.
i NaCl and NaBr *(3 marks)*
ii $FeSO_4$ and $Fe_2(SO_4)_3$ *(3 marks)*
iii NH_4Cl and $CrCl_3$ *(3 marks)*

Revision tip

Take care with this equation. Note that the balancing ensures that:

- total changes in oxidation number balance
- charges balance.

Notice that there are 4 I^- on the left but only two change oxidation number to I_2.

Revision tip

In a disproportionation reaction, the same element is reduced and oxidised.

Synoptic link

See Topic 8.2, The halogens, for other examples of disproportionation.

Synoptic link

For identification of anions (CO_3^{2-}, SO_4^{2-}, Cl^-, Br^-, I^-) and the cation NH_4^+, see Topic 8.3, Qualitative analysis.

Synoptic link

For identification of transition metal ions (Cu^{2+}, Fe^{2+}, Fe^{3+}, Mn^{2+}, Cr^{3+}), see Topic 24.4, Ligand substitution and precipitation.

Chapter 24 Practice questions

1 Which electron configuration is correct?

A Cu atom $1s^2 2s^2 2p^6 3s^2 3p^6 3d^9 4s^2$

B Cu atom $1s^2 2s^2 2p^6 3s^2 3p^6 3d^{10} 4s^1$

C Cu^+ ion $1s^2 2s^2 2p^6 3s^2 3p^6 3d^9 4s^1$

D Cu^+ ion $1s^2 2s^2 2p^6 3s^2 3p^6 3d^8 4s^2$ *(1 mark)*

2 Why is cobalt classified as a transition metal?

A Cobalt forms coloured compounds.

B Cobalt is a d-block element.

C Cobalt forms ions with incompletely filled d orbitals.

D Cobalt forms ions with different oxidation states. *(1 mark)*

3 Which complex ion has the metal with an oxidation number of +3?

A $[Co(NH_3)_4Cl_2]^+$ **B** VO_2^+

C $[Fe(CN)_6]^{4-}$ **D** $[Pt(NH_3)_3Cl]^+$ *(1 mark)*

4 a A complex ion of Fe^{3+} with four CN^- ligands and two NH_3 ligands exists as *cis* and *trans* stereoisomers.

i Write structural and empirical formulae for the complex ion. *(2 marks)*

ii Draw and label the *cis–trans* isomers of the complex ion. *(2 marks)*

b A complex ion of Cr^{3+} has two $H_2NCH_2CH_2NH_2$ ligands and two F^- ligands. This complex ion has *cis* and *trans* stereoisomers. One of the stereoisomers also shows optical isomerism.

i Write structural and empirical formulae for the complex ion. *(2 marks)*

ii Draw and label all the stereoisomers of the complex ion. *(3 marks)*

5 A complex of a transition metal **M** is dissolved in water forming a solution containing a complex ion **A**. Addition of NH_3(aq) forms a green precipitate, **M**$(OH)_2$, which did not dissolve in excess NH_3(aq). On standing in air, the green precipitate forms a red-brown precipitate **B**.

a Identify **M** and **A**. *(2 marks)*

b Write an equation for the formation of **M**$(OH)_2$ from the initial solution. *(1 mark)*

c Identify **B** and explain its formation. *(2 marks)*

6 A chromium(III) salt is dissolved in water forming a complex ion **A**. NaOH(aq) is added, forming a precipitate **B**. Addition of further NaOH(aq) forms a solution containing complex ion **C**.

a State the colours and formulae of **A**, **B**, and **C**. *(6 marks)*

b Write equations for

i the formation of **B** *(1 mark)*

ii the formation of **C** from **B** *(1 mark)*

iii the formation of **C** from **A**. *(1 mark)*

25.1 Introducing benzene

Specification reference: 6.1.1

The Kekulé and delocalised models for benzene

The Kekulé structure was the first model that was widely accepted for the structure of benzene, C_6H_6. The alternative delocalised model for benzene was developed later based on experimental evidence. Chemists use both representations in the formulae of aromatic compounds (See Figure 1).

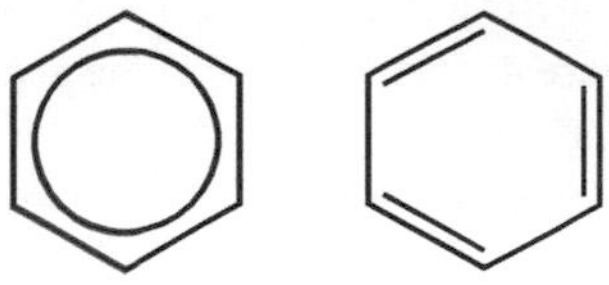

▲ **Figure 1** *Benzene in formulae*
- *The delocalised structure on the left.*
- *The Kekulé structure on the right.*

The Kekulé model

The Kekulé structure for benzene is a six-membered ring of carbon atoms joined by alternate C–C and C=C bonds. There are two ways of showing this arrangement, with the single and double bonds between different carbon atoms (Figure 2).

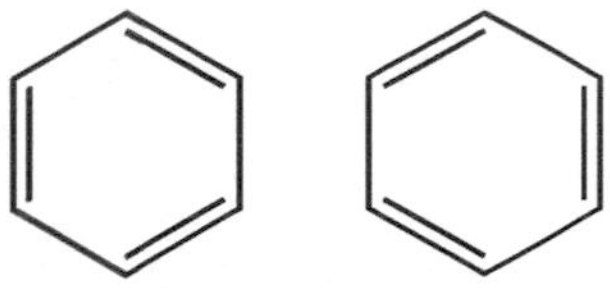

▲ **Figure 2** *Kekulé model of benzene*
Alternating C–C and C=C bonds can be drawn between different carbon atoms

The delocalised model of benzene

The delocalised structure of benzene treats all carbon–carbon bonds as being the same, somewhere between a single and double bond.

The main features of the delocalised model (Figure 3) are listed below.

- Benzene is a planar, cyclic, hexagonal hydrocarbon containing six C atoms and six H atoms.
- Each carbon atom has four electrons in its outer shell available for bonding.

  Three of the four electrons are used in σ-bonds (sigma bonds):
  - two σ-bonds to the C atoms on either side in the ring, C–C–C
  - one σ-bond to a H atom outside of the ring, C–H.
- Each carbon atom has one electron in a p-orbital, at right angles to the plane of the σ-bonded carbon and hydrogen atoms.

  Figure 3 shows how the p-orbitals overlap to form delocalised π-bonds.

Synoptic link

You learnt about the nature of the π-bond in alkenes in Topic 13.1, Properties of alkenes.

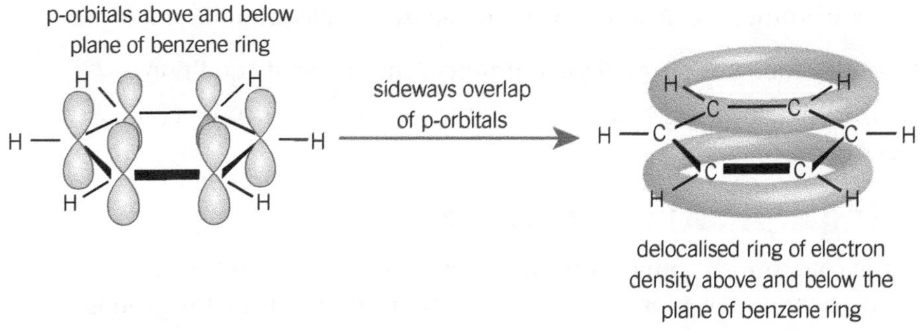

▲ **Figure 3** *The delocalised structure of benzene.*
The p-orbitals overlap sideways forming a π-electron cloud above and below the carbon ring

Experimental evidence for the delocalised model for benzene

Scientists used experimental evidence to develop the delocalised model of benzene as an improvement on the Kekulé model.

The key evidence is:

- bond lengths
- enthalpy change of hydrogenation
- resistance to reaction.

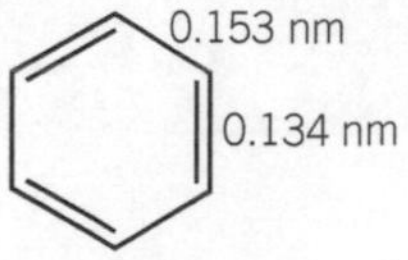

▲ **Figure 4** *Bond lengths in the Kekulé model of benzene*
In actual benzene, all bonds are the same length: 0.139 nm

Bond lengths

The Kekulé structure of benzene has single and double carbon–carbon bonds, which would have different bonds lengths: 0.153 nm for a single C–C bond, and 0.134 nm for a double C=C bond (Figure 4).

In benzene, all carbon–carbon bonds have the same length: 0.139 nm, between the bond lengths for C–C and C=C bonds.

This evidence suggests that all the carbon–carbon bonds are the same, somewhere between a single and a double carbon–carbon bond.

Enthalpy change of hydrogenation

Cyclohexene, with one C=C bond, has an enthalpy change of hydrogenation, ΔH, of $-120\,\text{kJ mol}^{-1}$ (Figure 5).

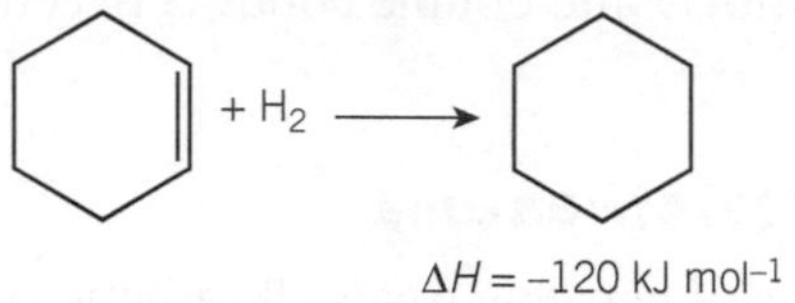

▲ **Figure 5** *Enthalpy change of hydrogenation of cyclohexene*

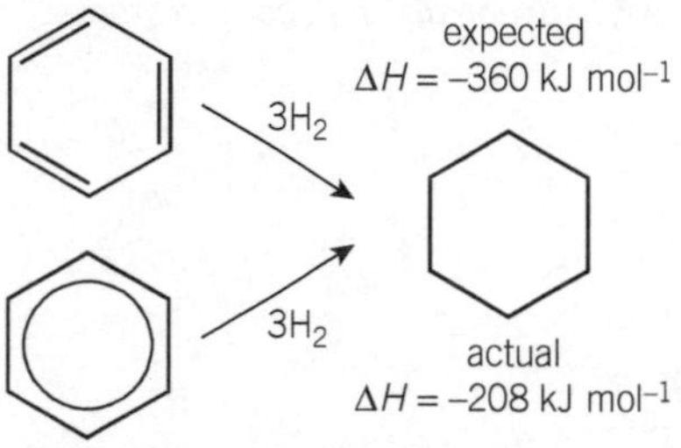

▲ **Figure 6** *Enthalpy changes of hydrogenation for benzene*

The Kekulé structure of benzene contains three C=C bonds.

For the enthalpy change of hydrogenation:

- expected ΔH for the Kekulé structure = $3 \times -120 = -360\,\text{kJ mol}^{-1}$
- experimental ΔH of benzene = $-208\,\text{kJ mol}^{-1}$.

The actual enthalpy change of hydrogenation of benzene is $152\,\text{kJ mol}^{-1}$ less exothermic than expected (Figure 6).

The actual structure of benzene is more stable than the Kekulé structure.

Synoptic link

Revisit Chapter 13, Alkenes, to revise the structure of alkenes.

Alkenes decolourise bromine water. See Topic 13.3, Reactions of alkenes.

Resistance to reaction

If benzene contained the C=C bonds in the Kekulé structure, it should decolourise bromine, in a similar way to alkenes by electrophilic addition.

- Benzene does **not** decolourise bromine under normal conditions
- Benzene is much less reactive than alkenes and does **not** react by electrophilic addition.

Naming aromatic compounds

Aromatic compounds contain substituent groups attached to the benzene ring in place of one or more hydrogen atoms. In their names, the groups are shown either as prefixes (before) to 'benzene' or suffixes (after) to 'phenyl', C_6H_5.

Revision tip

A **substituent group** is an atom, or group of atoms, taking the place of another atom, or group. In aromatic compounds based on benzene, a substituent group has replaced an H atom on the ring.

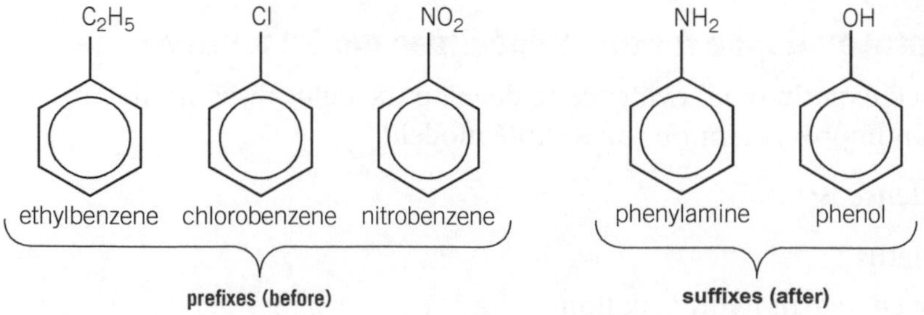

▲ **Figure 7** *Examples of prefixes and suffixes in naming*

Aromatic compounds with more than one substituent group

Some molecules may contain more than one substituent on the benzene ring (e.g. disubstituted compounds have two substituent groups).

The positions on the ring follow the same basic principles as for naming aliphatic compounds.

- The ring is numbered, just like a carbon chain, starting with one of the substituent groups.
- The positions of the groups on the ring are given the lowest possible numbers.
- The substituent groups are listed in alphabetical order.

Figures 8–9 show two examples of the names of disubstituted compounds.

Synoptic link

Revisit Topic 11.2, Nomenclature of organic compounds, if you need to revise the basics of naming aliphatic compounds.

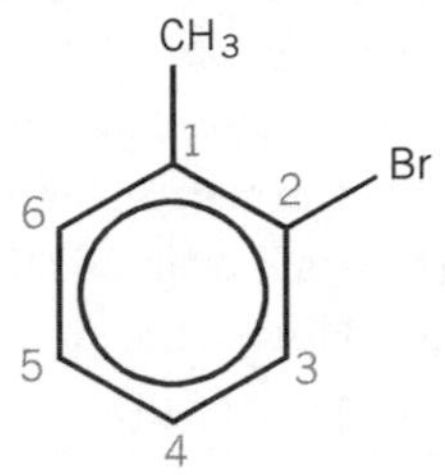

▲ **Figure 8** *2-bromomethylbenzene*

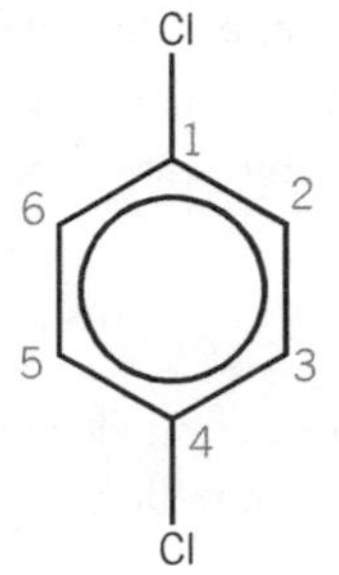

▲ **Figure 9** *1,4-dichlorobenzene*

Summary questions

1 a What does 'delocalised' mean? *(1 mark)*

b How do the following help to disprove the Kekulé model of benzene?

i bond lengths *(2 marks)*

ii reactivity *(2 marks)*

iii enthalpy change of hydrogenation. *(2 marks)*

2 Describe the pi-bonding in the delocalised model of benzene. *(3 marks)*

3 Draw the structures for the following aromatic compounds. *(3 marks)*

a 3-bromoethylbenzene

b 3,5-dinitrophenylamine

c 2,4-dibromo-6-chlorobenzoic acid

4 Name the following molecules. *(3 marks)*

a **b** **c**

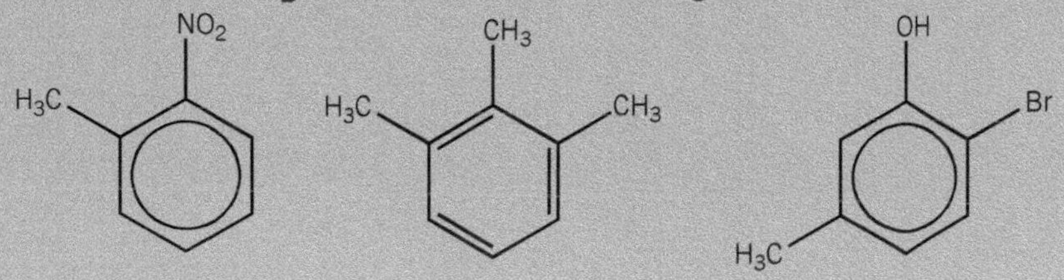

25.2 Electrophilic substitution reactions of benzene

Specification reference: 6.1.1

Synoptic link

You should recall the definitions of electrophile and substitution from Topic 13.4, Electrophilic addition in alkenes, and Topic 11.5, Introduction to reaction mechanisms.

Revision tip

In nitration, one of the hydrogen atoms on the benzene ring is substituted by a nitro, $-NO_2$, group.

Revision tip

NO_2^+ is the electrophile in the nitration of aromatic compounds.

Revision tip

The NO_2^+ electrophile accepts a pair of electrons from the benzene ring to form an unstable intermediate.

The intermediate then loses H^+ to form nitrobenzene. The stable benzene ring is reformed.

Electrophilic substitution in aromatic compounds

Benzene and substituted aromatic hydrocarbons (e.g. methylbenzene) are called **arenes**.

Arenes undergo substitution reactions in which a hydrogen atom on the benzene ring is replaced by an electrophile.

Nitration of benzene

Benzene reacts with concentrated nitric acid in the presence of concentrated sulfuric acid at 50 °C to form the substituted product, nitrobenzene, $C_6H_5NO_2$.

H_2SO_4 acts as a catalyst.

$$C_6H_6 + HNO_3 \rightarrow C_6H_5NO_2 + H_2O$$

Electrophilic substitution mechanism for nitration

The mechanism proceeds by three steps.

Step 1: Formation of the electrophile, NO_2^+

Concentrated HNO_3 and H_2SO_4 react to form the nitronium ion, NO_2^+.

$$HNO_3 + H_2SO_4 \rightarrow NO_2^+ + HSO_4^- + H_2O$$

nitronium ion

Step 2: Formation of the organic product, $C_6H_5NO_2$

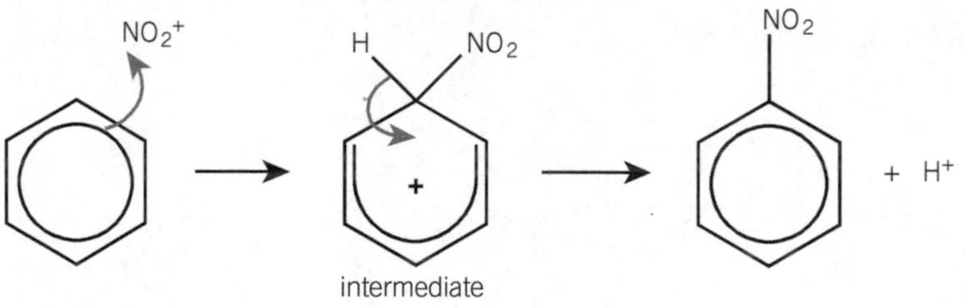

▲ **Figure 1** *The mechanism of benzene with nitric acid*

Step 3: Regeneration of the H_2SO_4 catalyst

The H^+ ion formed in **Step 2** reacts with the HSO_4^- ion from **Step 1** to regenerate the catalyst, H_2SO_4.

$$H^+ + HSO_4^- \rightarrow H_2SO_4$$

Revision tip

Halogen carriers are often generated *in situ* (in the reaction vessel) from the metal and the halogen (e.g. Fe and Br_2).

Revision tip

In bromination, one of the hydrogen atoms on the benzene ring is substituted by a bromine atom.

Halogenation of benzene

Halogens do not react with benzene unless a catalyst called a **halogen carrier** is present, e.g. $AlCl_3$, $FeCl_3$, $AlBr_3$, $FeBr_3$.

Bromination of benzene

Benzene reacts with bromine, in the presence of a halogen carrier, at room temperature to form the substituted product, bromobenzene, C_6H_5Br.

$$C_6H_6 + Br_2 \rightarrow C_6H_5Br + HBr$$

Electrophilic substitution mechanism for bromination

The electrophile is the bromonium ion, Br^+. The mechanism is similar to nitration and proceeds by three steps.

Step 1: Formation of the electrophile, Br^+

Bromine reacts with the halogen carrier to form the bromonium ion, Br^+.

$$Br_2 + FeBr_3 \rightarrow Br^+ + FeBr_4^-$$

bromonium ion

> **Revision tip**
> Br^+ is the electrophile in the bromination of aromatic compounds.

Step 2: Formation of the organic product, C_6H_5Br

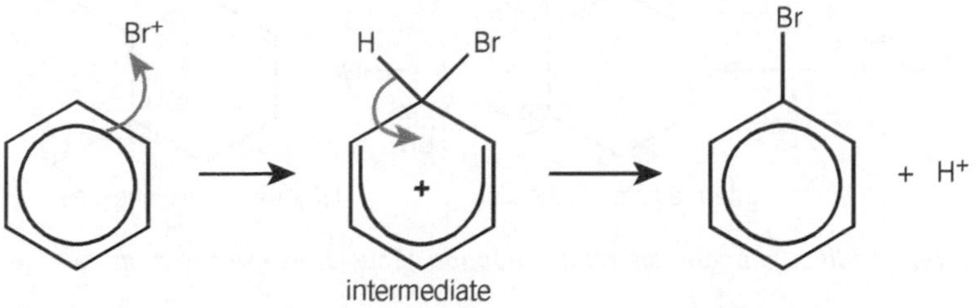

▲ **Figure 2** *The mechanism for the bromination of benzene*

> **Revision tip**
> The Br^+ electrophile accepts a pair of electrons from the benzene ring to form an unstable intermediate.
>
> The intermediate then loses H^+ to form bromobenzene. The stable benzene ring is reformed.

Step 3: Regeneration of the $FeBr_3$ catalyst

The H^+ ion formed in **Step 2** reacts with the $FeBr_4^-$ ion from **Step 1** to regenerate the catalyst, $FeBr_3$.

$$H^+ + FeBr_4^- \rightarrow FeBr_3 + HBr$$

Chlorination of benzene

Chlorine reacts with benzene to form chlorobenzene C_6H_5Cl. The reaction has the same mechanism as bromination, with a halogen carrier of $FeCl_3$, $AlCl_3$, or Fe (which forms $FeCl_3$ with Cl_2).

$$C_6H_6 + Cl_2 \rightarrow C_6H_5Cl + HCl$$

Alkylation and acylation of benzene (Friedel–Crafts reaction)

Alkylation and acylation proceed by electrophilic substitution and require the presence of a halogen carrier.

> **Synoptic link**
> Alkylation and acylation add a carbon chain to the benzene ring. See also, Topic 28.1, Carbon–carbon bond formation.

Alkylation

In alkylation, a haloalkane, e.g. RCl, is reacted with an aromatic compound to introduce an alkyl group to the benzene ring. A halogen carrier is required.

The equation shows the alkylation of benzene by chloroethane, C_2H_5Cl.

$$C_6H_6 + C_2H_5Cl \rightarrow C_6H_5Cl + HCl$$

> **Revision tip**
> In alkylation, one of the hydrogen atoms on the benzene ring is substituted by the alkyl group, R.

Acylation reactions

In acylation, an acyl chloride, RCOCl, is reacted with an aromatic compound to form an aromatic ketone. A halogen carrier is required.

Figure 3 shows the acylation of benzene by ethanoyl chloride, CH_3COCl.

> **Synoptic link**
> Acyl chlorides are discussed in more detail in Topic 26.4, Carboxylic acid derivatives.

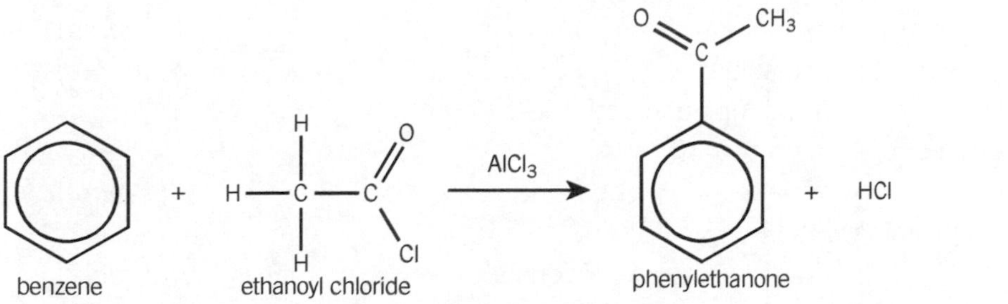

▲ **Figure 3** *The reaction between ethanoyl chloride and benzene, forming phenylethanone*

> **Revision tip**
> In acylation, one of the hydrogen atoms on the benzene ring is substituted by the acyl group, RCO.

Comparing the reactivity of alkenes with arenes

Alkenes react by **electrophilic addition**.

Arenes are less reactive and react by **electrophilic substitution**.

Electrophilic addition in alkenes

Alkenes, such as cyclohexene, react readily with bromine by addition.

$$C_6H_{10} + Br_2 \rightarrow C_6H_{10}Br_2$$

The mechanism for this reaction is **electrophilic addition** (Figure 5).

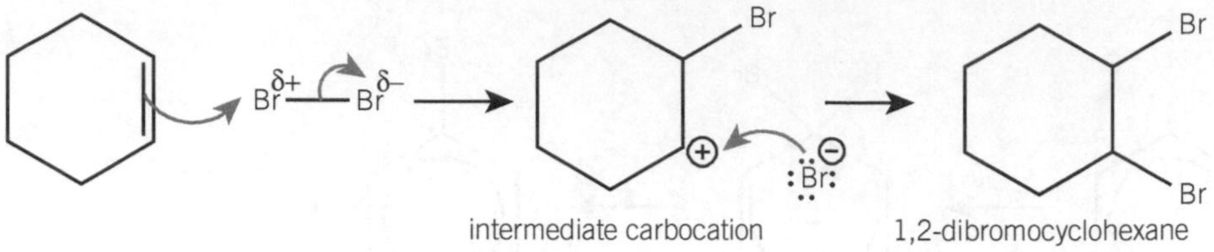

▲ **Figure 4** *Electrophilic addition mechanism for the reaction of cyclohexene and bromine*

Electrophilic substitution in benzene

Bromine is much less reactive with benzene than with alkenes. Benzene does react with bromine but only when a halogen carrier catalyst is present.

This reaction is **electrophilic substitution.** The mechanism is shown earlier in this topic.

- Benzene has delocalised π-electrons spread above and below the plane of the C atoms in the ring structure.
- The electron density around any two C atoms in the benzene ring is less than in the localised C=C double bond in an alkene.
- When the non-polar Br_2 molecule approaches the benzene ring, there is insufficient π-electron density around any two C atoms to polarise the bromine molecule.
  This prevents any reaction taking place without formation of Br^+ in the presence of a halogen carrier.

Revision tip

The relative resistance to bromination of benzene, compared with alkenes, was one of the key pieces of evidence that led to the delocalised model of benzene. For details, see Topic 25.1, Introducing benzene.

Synoptic link

The mechanism of electrophilic addition is covered in detail in Topic 13.4, Electrophilic addition of alkenes.

Revision tip

- Alkenes react with bromine by electrophilic addition.
- Benzene reacts with bromine by electrophilic substitution.

Revision tip

In alkenes, the π-bond has localised electrons above and below the plane of the C=C bond.

This produces an area of high electron density which readily attracts and polarises Br_2 molecules.

Summary questions

1 Benzene reacts by nitration and bromination in the presence of a catalyst.

a Name the mechanism. *(1 mark)*

b State the catalyst and write an equation for:

i nitration *(2 marks)* **ii** bromination. *(2 marks)*

2 Methylbenzene reacts with concentrated nitric acid to form a product, substituted at the 4-position.

a Name the organic product. *(1 mark)*

b Outline the mechanism, including stages involving the catalyst. *(4 marks)*

3 Phenylethanone, $C_6H_5COCH_3$ can be prepared from benzene.

a What reagents are needed for this preparation? *(2 marks)*

b Outline the mechanism, including the role of the catalyst. *(4 marks)*

4 Bromine reacts with propene and with benzene but the type of reaction is different.

a Name the mechanism for the reaction of bromine with

i propene *(1 mark)* **ii** benzene. *(1 mark)*

b Write an equation for the reaction of bromine with

i propene *(1 mark)* **ii** benzene. *(1 mark)*

c Explain why bromine reacts far more readily with propene than with benzene. *(3 marks)*

25.3 The chemistry of phenol

Specification reference: 6.1.1

Phenols

Phenols contain a hydroxyl, –OH, functional group bonded directly to an aromatic ring. Alcohols have the –OH group bonded to a carbon chain. See Figure 1. Although alcohols and phenols both have an –OH group, many reactions of phenols are different from alcohols.

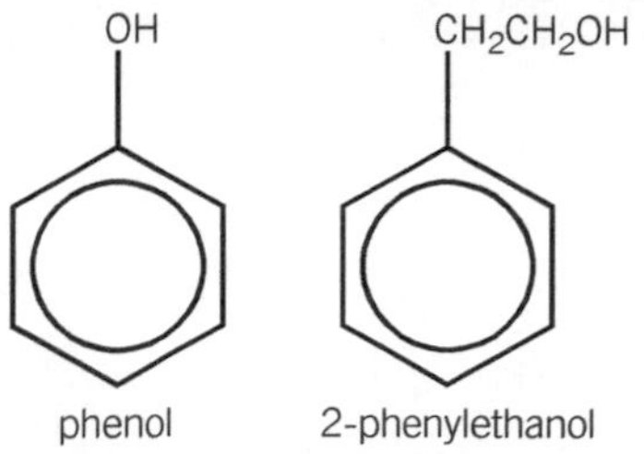

▲ **Figure 1** *Phenol, C_6H_5OH, (a phenol) and 2-phenylethanol (an alcohol)*

The acidity of phenol

Phenols as weak acids

Phenol are weak acids, partially dissociating in water to form the phenoxide ion and a proton (Figure 2). All phenols act as weak acids and turn pH paper an acidic colour.

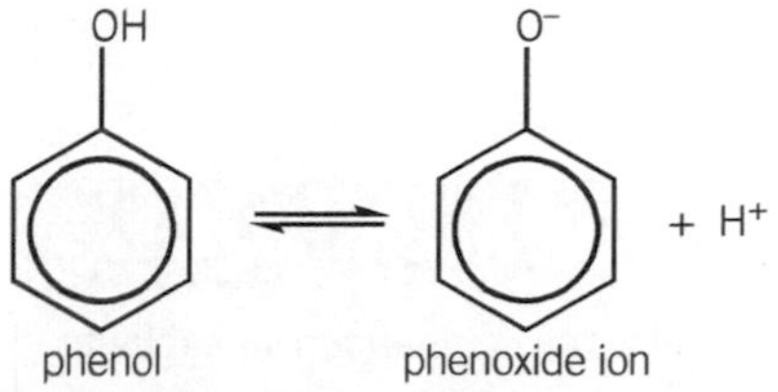

▲ **Figure 2** *Phenol as a weak acid*

Phenols are more acidic than alcohols but less acidic than carboxylic acids.

Sodium carbonate can be used to distinguish between a phenol and a carboxylic acid.

- Carboxylic acids react with Na_2CO_3 to form gas bubbles of carbon dioxide.
- Phenols do not react with Na_2CO_3 and there are no gas bubbles.

Synoptic link

For details of tests to distinguish organic functional groups, see Topic 29.1, Chromatography and functional group analysis.

Reaction of phenol with sodium hydroxide

Phenol reacts with sodium hydroxide to form the salt sodium phenoxide (C_6H_5ONa) and water in a neutralisation reaction (Figure 3).

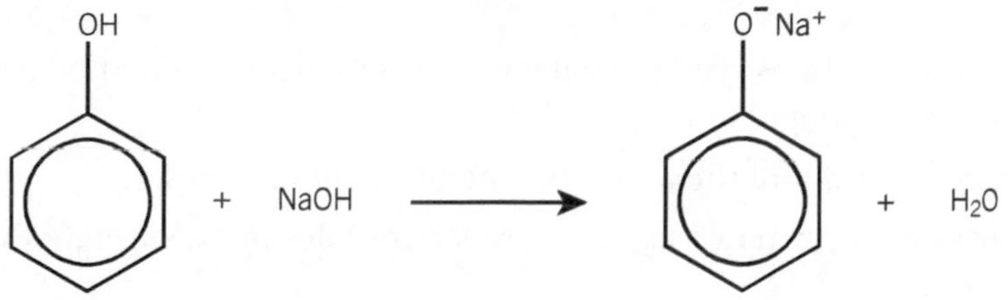

▲ **Figure 3** *The reaction of phenol with aqueous sodium hydroxide*

Electrophilic substitution reactions of phenol

The aromatic ring in phenols reacts by electrophilic substitution. Phenols are more reactive than benzene and the electrophilic substitution reactions take place under milder conditions.

Bromination of phenol

Phenol reacts with an aqueous solution of bromine (bromine water).

- The reaction takes place at room temperature and no halogen carrier is needed.
- Phenol decolourises the bromine and a white precipitate of 2,4,6-tribromophenol is formed.

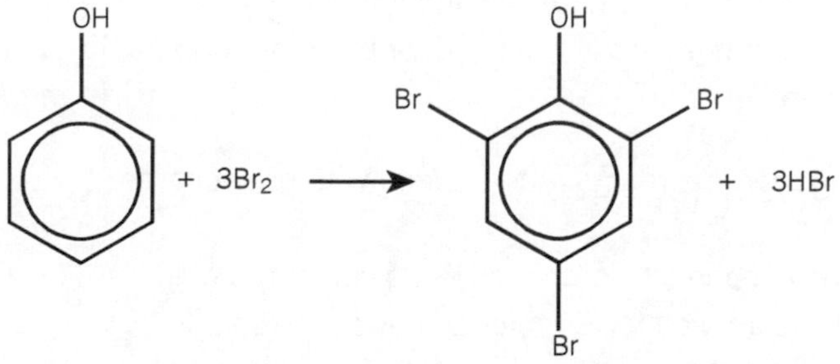

▲ **Figure 4** *The bromination of phenol produces 2,4,6-tribromophenol*

Synoptic link

The bromination and nitration of benzene are discussed in Topic 25.2, Electrophilic substitution reactions of benzene.

Nitration of phenol

Phenol reacts with dilute nitric acid to form a mixture of 2-nitrophenol and 4-nitrophenol. The equation to form 2-nitrophenol is shown below.

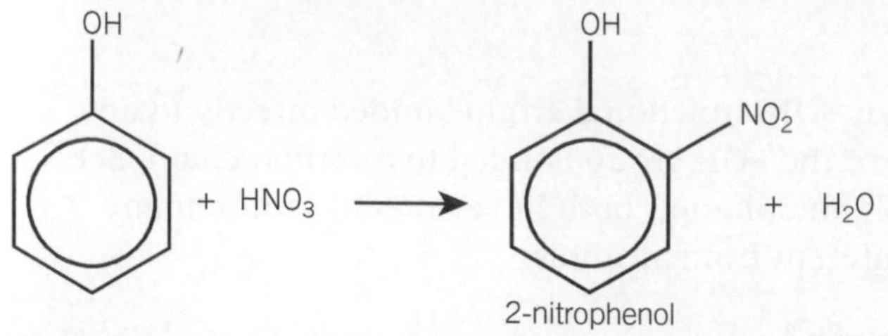

▲ **Figure 5** *Nitration of phenol to form 2-nitrophenol*

Comparing the reactivity of phenol and benzene

Bromine and nitric acid react more readily with phenol than with benzene.

	Phenol	Benzene
Bromination	• Reacts with bromine water at room temperature. • A trisubstituted organic product is formed.	• Reacts with bromine only with a halogen carrier. • A monosubstituted organic product is formed.
Nitration	• Reacts with dilute nitric acid at room temperature.	• Reacts with concentrated nitric and sulfuric acids at 50 °C.

Synoptic link

See also the comparison in reactivity of bromine with benzene and alkenes in Topic 25.2, Electrophilic substitution reactions of benzene.

Reasons for increased reactivity

The increased reactivity is caused by interaction of the hydroxyl group with the aromatic ring.

- A lone pair from an oxygen p-orbital of the –OH group is donated into the π-system of phenol.
- The electron density of the aromatic ring in phenol increases.
- The increased electron density attracts electrophiles more strongly than with benzene.

The aromatic ring in phenol is therefore more susceptible to attack from electrophiles than in benzene. Bromine molecules are polarised by the electron density in the phenol ring structure and so no halogen carrier catalyst is required.

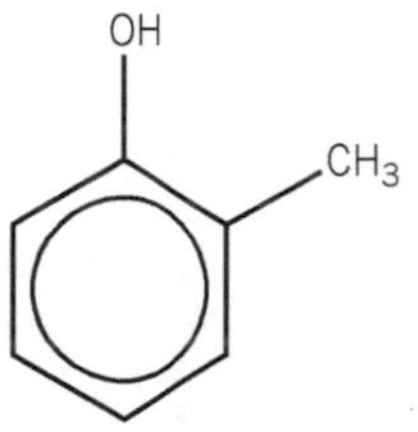

X

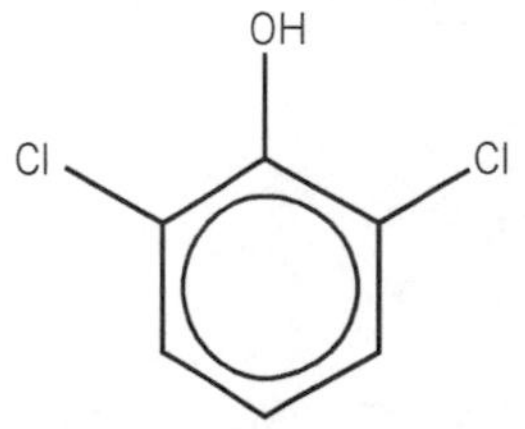

Y

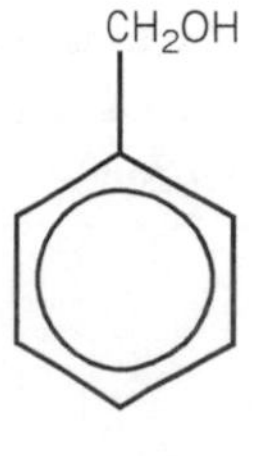

Z

Summary questions

1 **a** Name structures **X**, **Y**, and **Z**, and classify each as a phenol or an alcohol. *(3 marks)*

b What would you observe when phenol is tested with:

i pH indicator *(1 mark)* **ii** $Na_2CO_3(aq)$? *(1 mark)*

2 **a** Write an equation to show how phenol acts as a weak acid. *(1 mark)*

b Write an equation for the neutralisation of phenol. *(1 mark)*

c What different conditions are needed for bromination and nitration of benzene and phenol? *(4 marks)*

3 **a** Explain why bromine reacts more readily with phenol than with benzene. *(3 marks)*

b Compound **A** is a phenol. Compound **B** is a carboxylic acid. Compound **C** is an alcohol. Compound **D** is an alkene. How could you distinguish between **A**, **B**, **C**, and **D** using test-tube tests? *(4 marks)*

25.4 Disubstitution and directing groups

Specification reference: 6.1.1

Disubstitution

Benzene reacts by electrophilic substitution to form monosubstituted compounds, in which one hydrogen atom on the benzene ring is replaced by an electrophile.

The monosubstituted product can then undergo a second substitution – disubstitution. Further substitution can also take place forming multisubstituted compounds.

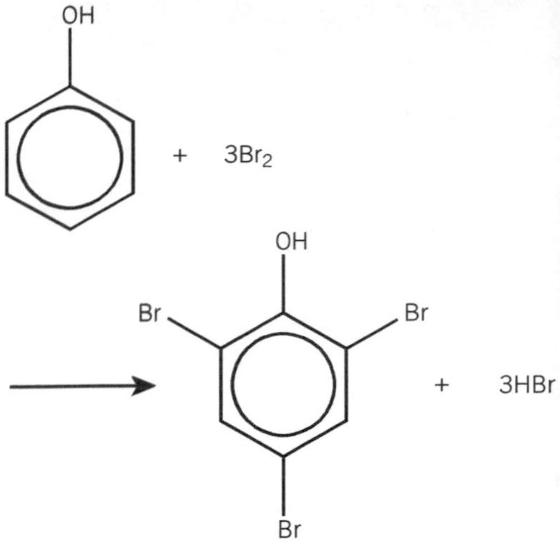

▲ **Figure 1** *The reaction of phenol with bromine produces 2,4,6-tribromophenol*

Directing effects

Disubstitution can take place more, or less, readily than benzene, depending on the first substituent group present. The position of the next substitution depends on whether the first substituent group is electron-donating or electron-withdrawing.

Electron-donating and electron-withdrawing groups

Benzene reacts with bromine to form bromobenzene. The reaction needs a halogen carrier.

$$C_6H_6 + Br_2 \rightarrow C_6H_5Br + HBr$$

Electron-donating groups

Phenol reacts with bromine to form 2,4,6-tribromophenol (Figure 1). No halogen carrier is required.

- The –OH group is electron-donating and increases the electron density of the aromatic ring.
- The aromatic ring of phenol is **activated** and reacts **more** readily with electrophiles than benzene does.

Revision tip

See Topic 25.3, The chemistry of phenol, for more details of the reaction of phenol with bromine and the reasons why phenol is more reactive than benzene.

Electron-withdrawing groups

Nitrobenzene reacts with bromine to form 3-bromonitrobenzene. The reaction requires both a halogen carrier and a high temperature.

- The $–NO_2$ group is electron-withdrawing and decreases the electron density of the aromatic ring.
- The ring is **deactivated** and reacts with electrophiles **less** readily than benzene does.

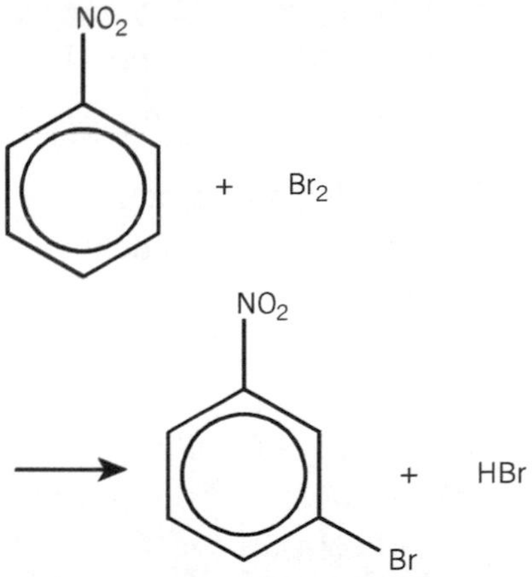

▲ **Figure 2** *The reaction of nitrobenzene with bromine forms 3-bromonitrobenzene*

Directing effects

The substitution position of a second group on the benzene ring depends on the directing effect of any groups already attached to the ring.

- Electron-donating groups are 2- and 4-directing.
- Electron-withdrawing groups (except for the halogens) are 3-directing.

Table 1 gives examples of different directing groups.

▼ **Table 1** *2,4-directing and 3-directing groups*

2,4-directing	3-directing
–OH	–CHO
$–NH_2$	–COR
–R (e.g. CH_3)	–COOH
–F, –Cl, –Br, –I	–COOR
	$–NO_2$
	–CN

Using directing groups in organic synthesis

Organic chemists use directing groups in the synthesis of compounds.

The Worked example shows how two different disubstituted aromatic compounds can be synthesised using the principles of directing groups.

Revision tip

You are expected to know the following directing effects:

- 2, 4-directing: $-OH$, $-NH_2$
- 3-directing: $-NO_2$

You are also expected to be able to predict disubstituted compounds from directing groups, provided in a table or otherwise.

Worked example: Synthesis of different isomers of methylnitrobenzene from benzene

From Table 1, the two substituents in methylnitrobenzene have different directing effects:

$-CH_3$: 2- and 4-directing

$-NO_2$: 3-directing.

- The different directing effects of $-CH_3$ and $-NO_2$ can be used to synthesise different isomers of methylnitrobenzene from benzene.
- The same reactions are used but in a different order.

Preparation of 3-methylnitrobenzene

Step 1: Nitration

React benzene with concentrated nitric and sulfuric acids at 50 °C.

The $-NO_2$ group is substituted onto the benzene ring.

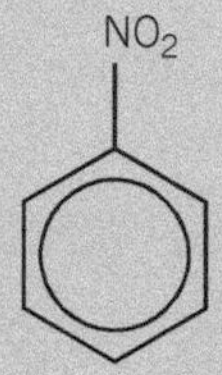

Step 2: Alkylation

React nitrobenzene with CH_3Br in the presence of a halogen carrier, $AlBr_3$.

The $-NO_2$ group is 3-directing and directs the $-CH_3$ group to the 3-position.

The product is 3-methylnitrobenzene.

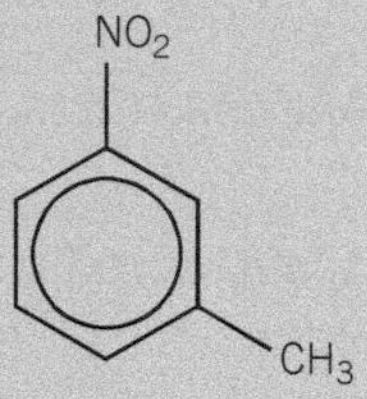

Preparation of 2-methylnitrobenzene and 4-methylnitrobenzene

Step 1: Alkylation

React benzene with CH_3Br in the presence of a halogen carrier, $AlBr_3$.

The product is 3-methylbenzene.

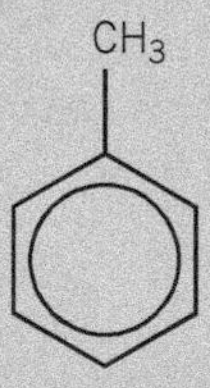

Step 2: Nitration

React with concentrated nitric and sulfuric acids at 50 °C.

The $-CH_3$ group is 2, 4-directing and directs the $-NO_2$ group to the 2- and 4-positions.

A mixture of 2-methylnitrobenzene and 4-methylnitrobenzene is formed. The isomers can then be separated.

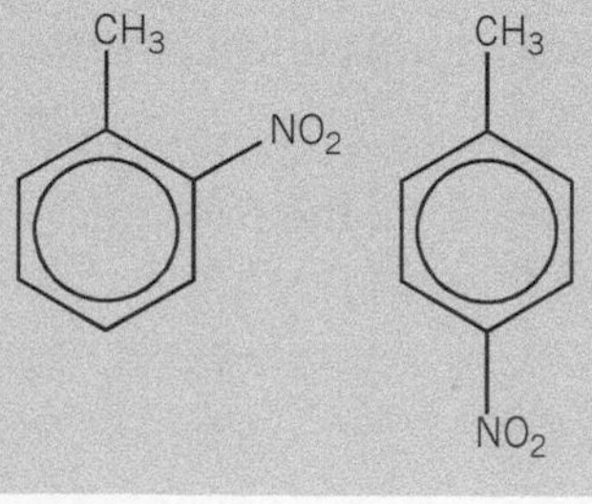

Summary questions

1 Identify the following compounds as containing 2,4-, or 3-directing groups. (Refer to Table 1.)

a C_6H_5I *(1 mark)*

b $C_6H_5NO_2$ *(1 mark)*

c C_6H_5CHO *(1 mark)*

d C_6H_5OH *(1 mark)*

2 Draw the organic products for the following reactions. (Refer to Table 1.)

a $C_6H_5OH + HNO_3$ *(1 mark)*

b $C_6H_5NO_2 + CH_3COCl$ (with $AlCl_3$) *(1 mark)*

c $C_6H_5CN + CH_3CH_2Cl$ (with $AlCl_3$) *(1 mark)*

3 Outline a two-stage synthesis from benzene for the following. (Refer to Table 1.) Include structures and reagents for each stage.

a 3-bromonitrobenzene *(2 marks)*

b 4-bromonitrobenzene *(2 marks)*

Chapter 25 Practice questions

1 How many arene isomers have the molecular formula C_8H_{10}?

A 1 **B** 2 **C** 3 **D** 4 *(1 mark)*

2 Which statement is **not** correct for the delocalised model of benzene?

A Benzene reacts with bromine, only with a halogen carrier.

B Benzene has carbon–carbon bonds that all have the same length.

C Benzene has planar molecules.

D Benzene molecules are alicyclic. *(1 mark)*

3 Which statement describes part of the mechanism for the nitration of benzene?

A An intermediate donates a pair of electrons.

B The intermediate is a carbanion.

C The benzene ring donates a pair of electrons.

D The nitronium ion is a nucleophile. *(1 mark)*

4 Which statement correctly describes the –OH group in phenol?

A The –OH group is 2,4-directing and electron-withdrawing.

B The –OH group is 3-directing and electron-withdrawing.

C The –OH group is 2,4-directing and electron-donating.

D The –OH group is 3-directing and electron-donating. *(1 mark)*

5 Bromine is reacted with cyclohexene and phenol.

a Name the organic product and mechanism for the reaction of bromine with

i cyclohexene *(2 marks)*

ii phenol. *(2 marks)*

b Benzene reacts with bromine only in the presence of a halogen carrier.

i State a suitable halogen carrier. *(1 mark)*

ii Outline the mechanism for the bromination of benzene. *(5 marks)*

c Explain why bromine reacts less readily with benzene than with phenol. *(3 marks)*

6 Phenol, C_6H_5OH, is a weak acid.

a Write an equation to show the dissociation of phenol. *(1 mark)*

b Phenol, C_6H_5OH, is reacted with NaOH(aq).

i Write an equation for the reaction and name the type of reaction. *(1 mark)*

ii Name the organic product formed and the type of reaction. *(2 marks)*

c When magnesium is added to a solution of phenol, a gas is produced.

Predict the equation and name the type of reaction. *(3 marks)*

d Phenol reacts with dilute nitric acid to form a mixture of organic products.

i Name the organic products. *(2 marks)*

ii Write an equation for the formation of one of the organic products. *(2 marks)*

iii Predict the organic product from reacting phenol with concentrated HNO_3 and H_2SO_4.

Name the organic product and draw its structure. *(2 marks)*

26.1 Carbonyl compounds

Specification reference: 6.1.2

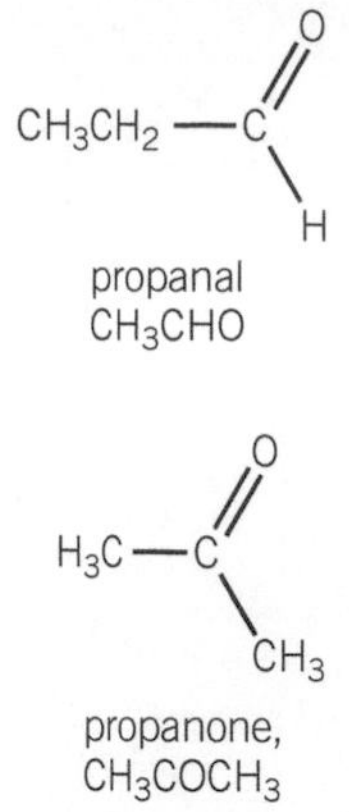

▲ **Figure 1** *An aldehyde (propanal) and a ketone (propanone)*

Carbonyl compounds

Aldehydes and ketones are organic compounds containing the carbonyl functional group, C=O.

Aldehydes

In aldehydes, the carbonyl functional group is at the end of a carbon chain. In a structural formula, the aldehyde group is written as CHO, e.g. CH_3CH_2CHO, propanal ('-al' for aldehyde).

Ketones

In ketones, the carbonyl functional group is between two carbon atoms in the carbon chain. In its structural formula, the ketone group is written as CO, e.g. CH_3COCH_3, propanone ('-one' for ketone).

Figure 1 shows the structures of propanal and propanone.

Oxidation of carbonyl compounds

Aldehydes

Aldehydes, RCHO, are oxidised to carboxylic acids, RCOOH, by refluxing with acidified dichromate(VI) ions, $H^+(aq)/Cr_2O_7^{2-}(aq)$ (the oxidising agent).

- H_2SO_4 and $K_2Cr_2O_7$ (or $Na_2Cr_2O_7$) can be used as a source of $H^+(aq)/Cr_2O_7^{2-}(aq)$.

Figure 2 shows the oxidation of butanal to butanoic acid by $H^+(aq)/Cr_2O_7^{2-}(aq)$.

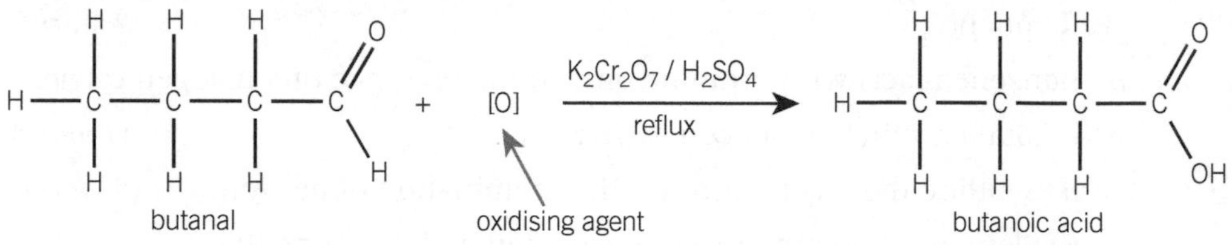

$CH_3CH_2CH_2CHO$ + [O] → $CH_3CH_2CH_2COOH$

aldehyde → **carboxylic acid**

▲ **Figure 2** *The oxidation of butanal with excess acidified potassium dichromate forming butanoic acid*

Ketones

Ketones do not undergo oxidation reactions. This lack of reactivity provides a way of distinguishing between aldehydes and ketones. An unknown carbonyl compound can be heated with $H^+(aq)/Cr_2O_7^{2-}(aq)$. If the colour changes from orange to green, the compound is an aldehyde.

Nucleophilic addition reactions of the carbonyl group

The carbonyl functional group is polar: $^{\delta+}C=O^{\delta-}$.

Due to the polarity of the C=O double bond, aldehydes and ketones react with some nucleophiles.

- The electron-deficient carbon atom in the C=O bond attracts nucleophiles.
- Addition occurs across the C=O double bond.

The reaction type is **nucleophilic addition**.

Synoptic link

Look back at Topic 14.2, Reactions of alcohols, to revise how primary alcohols can be oxidised to form aldehydes and carboxylic acids, and how secondary alcohols can be oxidised to ketones.

Revision tip

In equations, [O] is used to represent the oxidising agent, $H_2SO_4/K_2Cr_2O_7$. In the reaction, orange $Cr_2O_7^{2-}$ ions are reduced to green Cr^{3+} ions.

Synoptic link

See Topic 24.5, Redox and qualitative analysis, for details of the reduction of $Cr_2O_7^{2-}$ to Cr^{3+}.

Synoptic link

In Topic 26.2, Identifying aldehydes and ketones, you will learn more about the chemical tests that can be used to show the presence of a carbonyl group, and to distinguish aldehydes from ketones.

Synoptic link

You will recall from Topic 15.1, The chemistry of the haloalkanes, that a nucleophile donates an electron pair to an electron-deficient carbon atom, to form a new covalent bond.

Reaction of carbonyl compounds with $NaBH_4$

Aldehydes and ketones are reduced to alcohols by the reducing agent, sodium tetrahydridoborate(III), $NaBH_4$.

Reduction of an aldehyde

Aldehydes are reduced by $NaBH_4$ to **primary alcohols.**

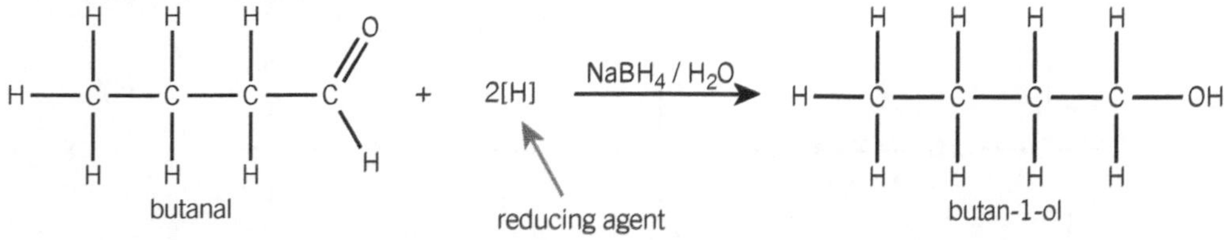

$CH_3CH_2CH_2CHO$ + 2[H] → $CH_3CH_2CH_2CH_2OH$

aldehyde → **primary alcohol**

▲ **Figure 3** *Butanal is reduced to butan-1-ol, a primary alcohol*

> **Revision tip**
> In the equation, [H] is used to represent the reducing agent, $NaBH_4$.

Reduction of a ketone

Ketones are reduced by $NaBH_4$ to **secondary alcohols.**

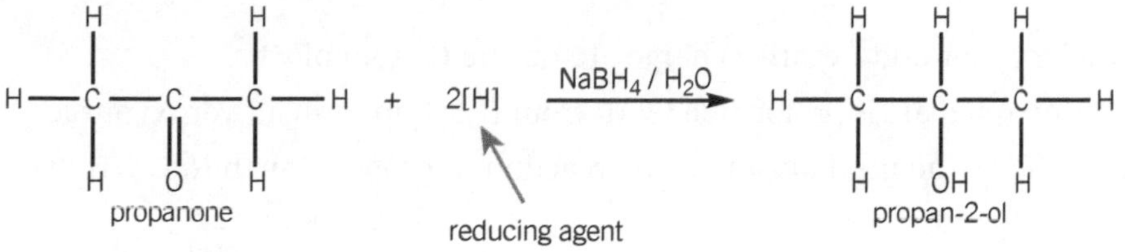

CH_3COCH_3 + 2[H] → $CH_3CHOHCH_3$

ketone → **secondary alcohol**

▲ **Figure 4** *Propanone is reduced to propan-2-ol, a secondary alcohol*

Reaction of carbonyl compounds with HCN

Hydrogen cyanide, HCN, adds across the C=O bond of aldehydes and ketones to form a hydroxynitrile (containing –OH and –CN functional groups). HCN is a very poisonous gas and HCN is generated in solution using sodium cyanide, NaCN and sulfuric acid, H_2SO_4.

> **Revision tip**
> The reactions of carbonyl compounds with $NaBH_4$ and HCN are addition reactions as two reactants combine to become one product.

The reaction of propanal with hydrogen cyanide is shown in Figure 5.

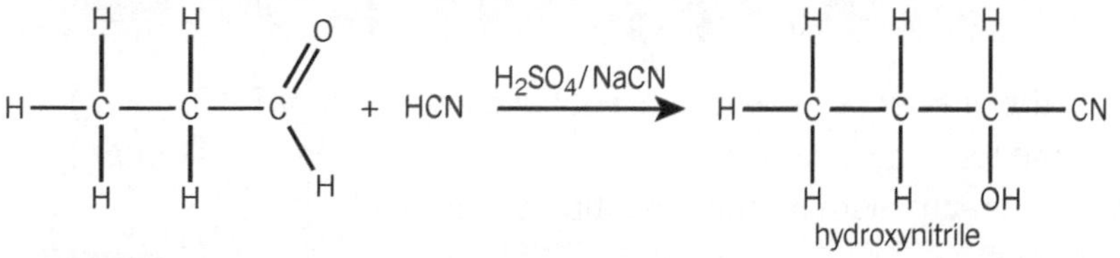

▲ **Figure 5** *Propanal reacts with hydrogen cyanide to form a hydroxynitrile*

> **Revision tip**
> The –CN functional group is called a nitrile. The old name for CN is 'cyanide', which is used for 'hydrogen cyanide', HCN.

> **Synoptic link**
> Using HCN to extend a carbon chain is covered in detail in Topic 28.1, Carbon-carbon bond formation.

The reaction is very useful in organic synthesis as it extends the length of the carbon chain. In this reaction, the three-carbon chain in propanal is lengthened to a four-carbon chain in the hydroxynitrile product.

Mechanism for nucleophilic addition to carbonyl compounds

Mechanism for the reaction with $NaBH_4$

$NaBH_4$ can be considered to contain the hydride ion, $:H^-$, which acts as the nucleophile.

Step 1: Nucleophilic attack → Intermediate.

- The H^- nucleophile is attracted to the electron-deficient carbon atom in the C=O bond.
- H^- bonds to the $^{\delta+}C$ atom.
- The C=O bond breaks to form a negatively charged intermediate.

Step 2: Protonation → alcohol product.

- The O^- atom of the intermediate is protonated by H_2O to form an alcohol.

Overall, hydrogen has been added across the C=O double bond.

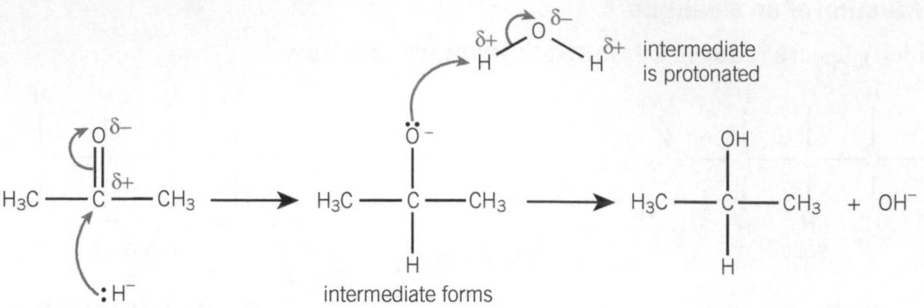

▲ **Figure 6** *Reduction of a carbonyl compound by nucleophilic addition*

Revision tip

Nucleophilic addition is one of the mechanisms that you are required to learn. The mechanism applies to the C=O group in aldehydes and ketones.

Ensure that curly arrows are positioned precisely and that all charges and partial charges have been included.

The mechanism for the reaction with NaCN/H⁺

In this reaction, the nucleophile is the cyanide ion, $:CN^-$ (from NaCN).

The mechanism is essentially the same as for H^- from $NaBH_4$ but note the special points below:

- The lone pair and negative charge are on the C atom of CN^-.
- The intermediate is protonated by H^+ from H_2SO_4 to form a hydroxynitrile.

Figure 7 shows the mechanism for the reaction of propanal with $NaCN/H_2SO_4$

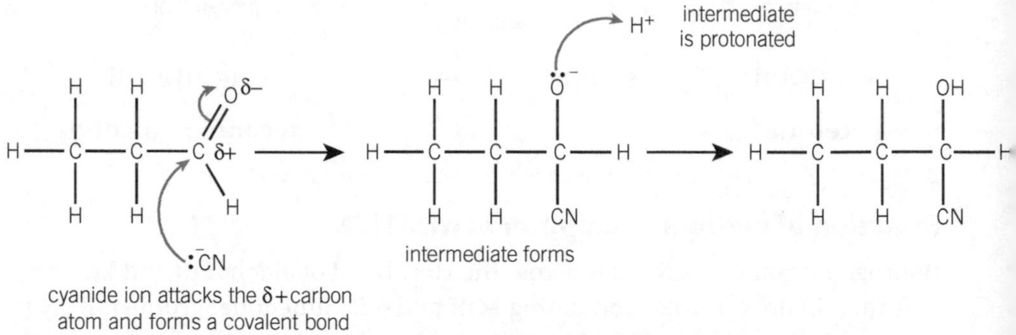

▲ **Figure 7** *Reaction of propanal with cyanide ions to form a hydroxynitrile*

Revision tip

In the first stage of the mechanism, be very careful that you show the lone pair and '–' charge on the C atom of the CN^- ion.

It is unfortunate that a cyanide ion is normally written in text as CN^-.

Summary questions

1 a What is the difference between an aldehyde and a ketone? *(2 marks)*
b Define the term 'nucleophile'. *(1 mark)*
c Name the mechanism for the reduction of carbonyl compounds. *(1 mark)*

2 a State the reagents and write equations for the following reactions of ethanal.
i Oxidation. *(2 marks)*
ii Reaction with HCN. *(2 marks)*
b State the reagents for the reduction of carbonyl compounds. *(1 mark)*
c Write an equation for the reduction of:
i pentanal *(1 mark)*
ii butanone. *(1 mark)*

3 Outline the mechanism for the reaction between propanone and $NaCN/H_2SO_4$. *(4 marks)*

26.2 Identifying aldehydes and ketones

Specification reference: 6.1.2

Detecting a carbonyl group, C=O

The carbonyl C=O group in aldehydes and ketones can be detected using the following test.

- The compound is added to a solution of **2,4-dinitrophenylhydrazine** (2,4-DNP).
- In the presence of a C=O group, a yellow/orange precipitate forms.

Identification from the melting point of the 2,4-DNP derivative

The yellow/orange precipitate (the 2,4-DNP derivative) can be analysed to identify the carbonyl compound as follows.

- The impure yellow/orange solid is filtered and then recrystallised to produce a pure sample of the 2,4-DNP derivative.
- The melting point of the purified 2,4-DNP derivative is determined.
- The melting point is matched to a database of melting points for 2,4-DNP derivatives of carbonyl compounds. See Table 1 for an example.

Detecting an aldehyde group, CHO

The aldehyde –CHO group can be detected using the following test.

- The compound is warmed with **Tollens' reagent**.
- In the presence of an aldehyde –CHO group, a silver mirror forms.

Distinguishing between aldehydes and ketones

Aldehydes and ketones can easily be distinguished using Tollens' and 2,4-DNP tests.

- Only aldehydes form a silver mirror with Tollens' reagent.
- A ketone does **not** form silver mirror with Tollens' reagent, but does form a yellow/orange precipitate with 2,4-DNP.

Reaction of aldehydes with Tollens' reagent

Tollens' reagent contains silver(I) ions, $Ag^+(aq)$, in aqueous ammonia.

- An aldehyde is oxidised by the Ag^+ ions to a carboxylic acid:

  $RCHO + [O] \rightarrow RCOOH$

- The Ag^+ ions are reduced to form silver which shows up as a silver mirror:

  $Ag^+(aq) + e^- \rightarrow Ag(s)$

Ketones cannot be oxidised and do **not** form a silver mirror with Tollens' reagent.

Summary questions

1 a State the observations and conclusions from a positive test with 2,4-DNP. *(2 marks)*

b State the observations and conclusions from a positive test with Tollens' reagent. *(2 marks)*

2 a Write oxidation and reduction equations for a positive test with Tollens' reagent. *(2 marks)*

b Describe how a carbonyl compound can be identified. *(3 marks)*

3 You are provided with the following compounds:

CH_3CH_2CHO $(CH_3)_2CHCOCH_3$ $CH_3CH_2CH_2CHO$ $CH_3CH_2CH_2CH_2OH$

How could you distinguish between these compounds? *(6 marks)*

Revision tip

The solution of 2,4-DNP used is known as Brady's Reagent.

You are not expected to know the structure of 2,4-DNP or the derivative formed in the test.

▼ **Table 1** *Melting points of 2,4-DNP derivatives of carbonyl compounds*

Carbonyl compound	Melting point of 2,4-DNP derivative/°C
pentanal	98
propanone	128
propanal	156

Synoptic link

You will cover recrystallisation and determination of melting points in Topic 28.2, Further practical techniques.

Revision tip

Tollens' reagent is sometimes known as 'ammoniacal silver nitrate'.

Revision tip

Instead of using Tollens' reagent, an unknown carbonyl compound could be warmed with acidified dichromate(VI). A colour change from orange to green confirms an aldehyde.

For details of the oxidation of aldehydes, see Topic 26.1, Carbonyl compounds.

Synoptic link

For further details of tests for organic functional groups, see Topic 29.1, Chromatography and functional group analysis.

26.3 Carboxylic acids

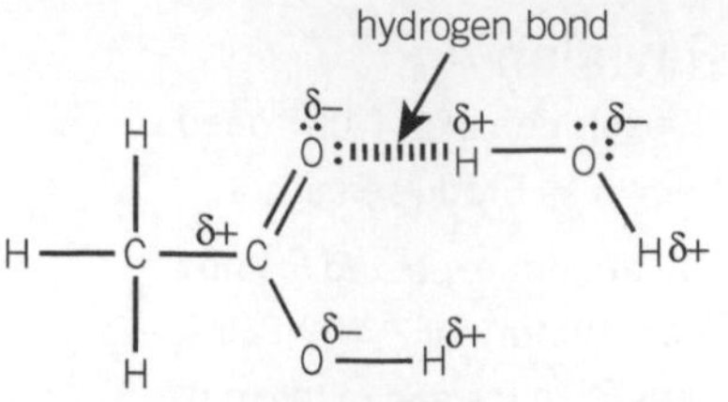

▲ **Figure 1** *Hydrogen bonding between carboxylic acid and water molecules*

Synoptic link

For details of hydrogen bonding, see Topic 6.4, Hydrogen bonding.

Synoptic link

For more details of redox reactions of acids with metals, see Topic 4.3, Redox, and Topic 20.1, Brønsted–Lowry acids and bases.

Synoptic link

For more details of the dissociation of weak acids and neutralisation reactions, see Topic 4.1, Acids, bases, and neutralisation.

Ionic equations can be written for all these 'acid' reactions and they match those of a typical acid. See Topic 20.1, Brønsted–Lowry acids and bases.

Synoptic link

For further details of tests for organic functional groups, see Topic 29.1, Chromatography and functional group analysis.

Carboxyl compounds

Carboxylic acids contain the carboxyl functional group, –COOH.

Water solubility of carboxylic acids

The C=O and O–H bonds in carboxylic acids are polar allowing carboxylic acids to form hydrogen bonds with water molecules (Figure 1).

The alkyl carbon chain is non-polar. Water solubility decreases as the carbon chain length increases and the non-polar alkyl group becomes more significant.

Acid reactions of carboxylic acids

Carboxylic acids are weak acids, only partially dissociating in water.

$$RCOOH(aq) \rightleftharpoons H^+(aq) + RCOO^-(aq)$$

Redox reactions of carboxylic acids with metals

In aqueous solution, carboxylic acids react with many metals in a redox reaction to form a carboxylate salt and hydrogen gas.

e.g. $2CH_3COOH(aq) + Mg(s) \rightarrow (CH_3COO^-)_2Mg^{2+}(aq) + H_2(g)$

Neutralisation reactions of carboxylic acids with bases

Aqueous solutions of carboxylic acids are neutralised by bases

Reaction with metal oxides

Carboxylic acids react with metal oxides to form a salt and water.

e.g. $2CH_3COOH(aq) + MgO(s) \rightarrow (CH_3COO^-)_2Mg^{2+}(aq) + H_2O(l)$

Reaction with alkalis

Carboxylic acids react with alkalis to form a salt and water.

e.g. $CH_3COOH(aq) + NaOH(aq) \rightarrow CH_3COO^-Na^+(aq) + H_2O(l)$

Reaction with carbonates

Carboxylic acids react with carbonates to form a salt, CO_2, and H_2O.

e.g. $2CH_3COOH(aq) + Na_2CO_3(aq) \rightarrow 2CH_3COO^-Na^+(aq) + CO_2(g) + H_2O(l)$

This reaction is used as a test for the COOH group.

Summary questions

1 **a** Why are carboxylic acids soluble in water? *(2 marks)*
b Using methanoic acid with an equation, explain what is meant by a weak acid. *(2 marks)*

2 Write equations, using structural formulae, for the reactions between:
a propanoic acid and calcium oxide *(1 mark)*
b methanoic acid and calcium carbonate *(1 mark)*
c butanoic acid with zinc. *(1 mark)*

3 25 cm^3 of 0.100 mol dm^{-3} ethanedioic acid, HOOC–COOH, is reacted with $Na_2CO_3(aq)$ and with NaOH(aq).
Write equations, using structural formulae, for the reactions with
a 25 cm^3 0.100 mol dm^{-3} Na_2CO_3 *(1 mark)*
b 25 cm^3 0.100 mol dm^{-3} NaOH. *(1 mark)*

26.4 Carboxylic acid derivatives

Specification reference: 6.1.3

Carboxylic acid derivatives

Carboxylic acid derivatives contain an acyl group (Figure 1). A derivative of a carboxylic acid can be hydrolysed to form the parent carboxylic acid.

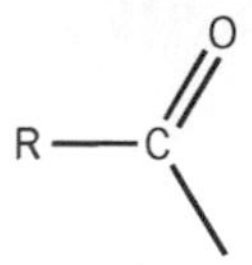

▲ **Figure 1** *The acyl group*

Figure 2 shows common derivatives of carboxylic acids.

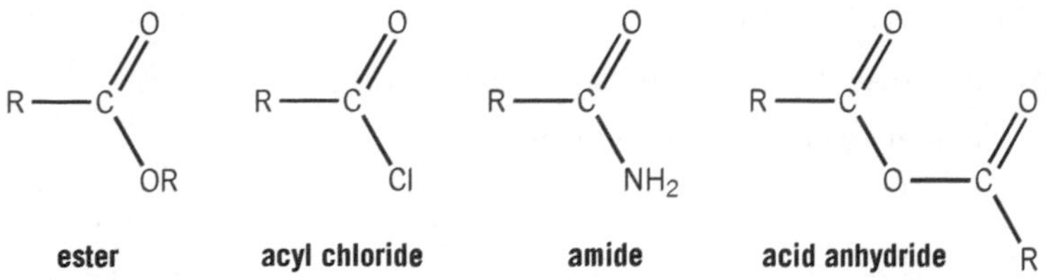

▲ **Figure 2** *Carboxylic acid derivatives*

Esters

Esters are common, naturally-occurring compounds. Most esters are sweet-smelling liquids and used in many perfumes and in food flavouring.

Naming esters

Naming of an ester is shown in Figure 3.

- The –oic acid suffix from the parent carboxylic acid is replaced with –oate.
- The alkyl chain attached to the COO group is then added as the first word in the name.

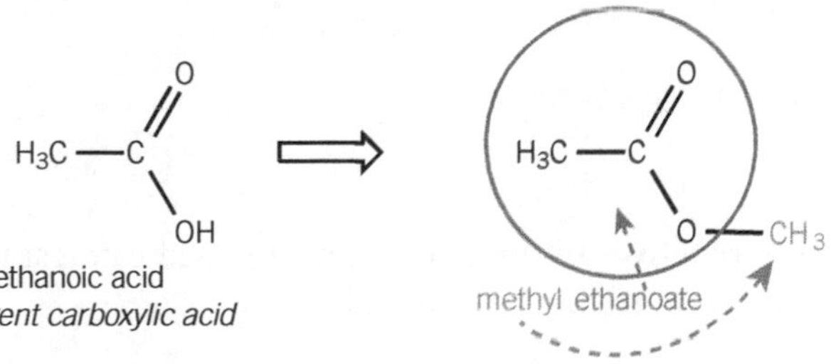

▲ **Figure 3** *Naming esters*

Revision tip

You will come across esters many times. Make sure that you can name esters and write their formulae.

Esterification

Esterification is the reaction of an alcohol with a carboxylic acid to form an ester.

- An alcohol is warmed with a carboxylic acid.
- A small amount of concentrated sulfuric acid is added, which acts as a catalyst.

Figure 4 shows the esterification of propanoic acid with methanol.

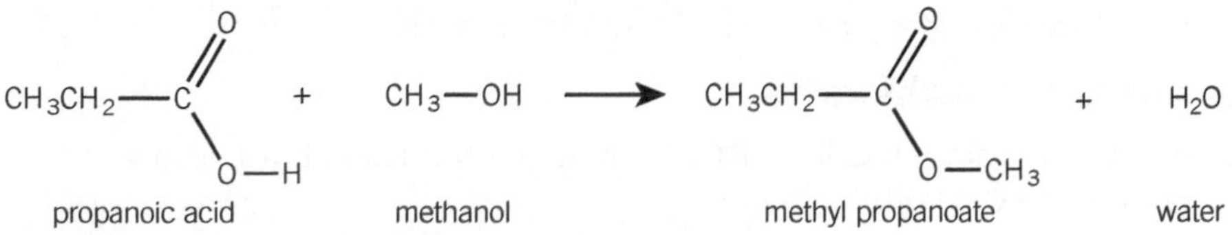

▲ **Figure 4** *The preparation of the ester methyl propanoate*

Revision tip

Notice how the name of the ester, methyl propanoate, has been derived from propanoic acid and methanol.

Hydrolysis of esters

Hydrolysis is the chemical breakdown of a compound in the presence of water or in aqueous solution.

Esters can be hydrolysed by hot aqueous acid or alkali to form the parent carboxylic acid or the carboxylate ion.

Acid hydrolysis of an ester

Acid hydrolysis of an ester is the reverse reaction of esterification.

- The ester is refluxed with dilute aqueous acid.
- The ester is broken down by water, with the acid acting as a catalyst.

Acid hydrolysis of an ester forms a carboxylic acid and an alcohol:

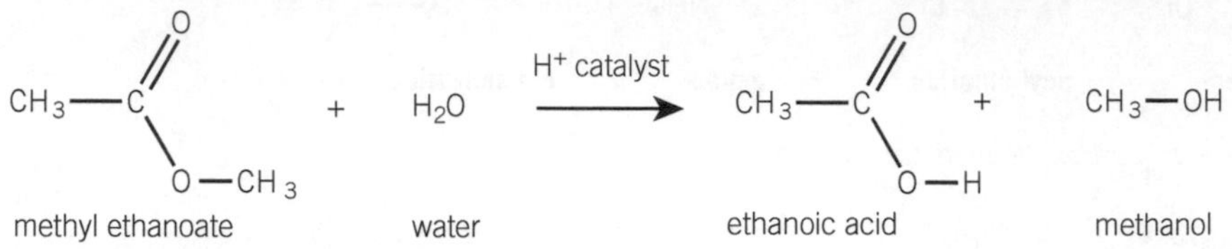

▲ **Figure 5** *The acid hydrolysis of methyl ethanoate*

Alkaline hydrolysis of an ester

Alkaline hydrolysis is irreversible. The ester is refluxed with aqueous alkali.

Alkaline hydrolysis of an ester forms a carboxylate ion and an alcohol:

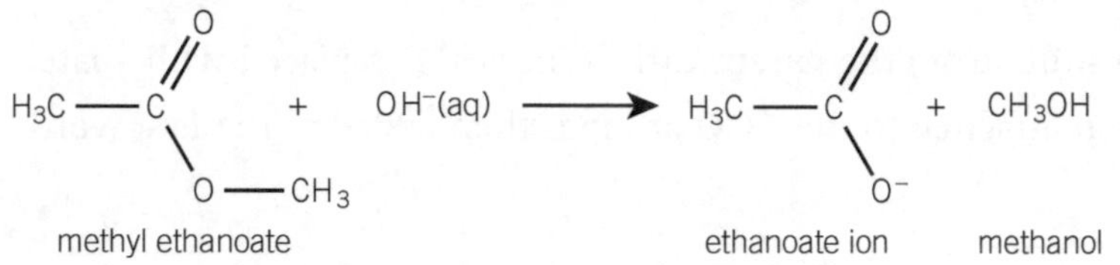

▲ **Figure 6** *The alkaline hydrolysis of methyl ethanoate*

Revision tip

If aqueous sodium hydroxide is the alkali, hydrolysis forms the carboxylate salt, sodium ethanoate:

$CH_3COOCH_3 + NaOH \rightarrow CH_3COO^-Na^+ + CH_3OH$

Acyl chlorides

Acyl chlorides, RCOCl, are very reactive. In their reactions, no acid catalyst is required and the reactions often produce very good yields.

Preparation of acyl chlorides

Acyl chlorides can be prepared by reacting a carboxylic acid with thionyl chloride, $SOCl_2$:

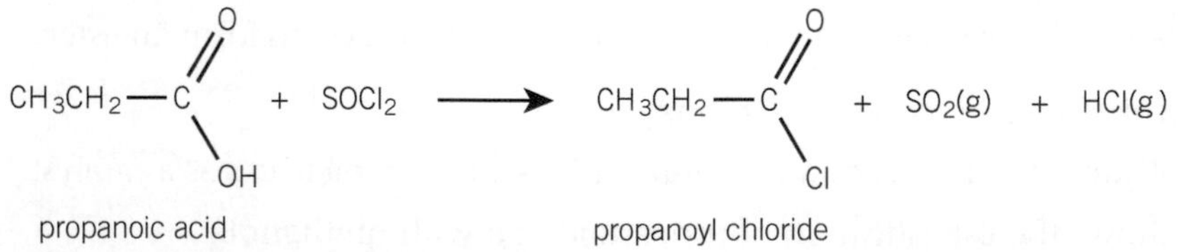

▲ **Figure 7** *The reaction of an acyl chloride with an alcohol forming an ester*

Revision tip

The other products of this reaction, SO_2 and HCl, are evolved as gases, leaving just the acyl chloride.

Reactions of acyl chlorides

In their reactions, acyl chlorides, RCOCl, react with nucleophiles, losing –Cl, but retaining the C=O double bond.

Figure 8 shows an ester forming from an acyl chloride and an alcohol.

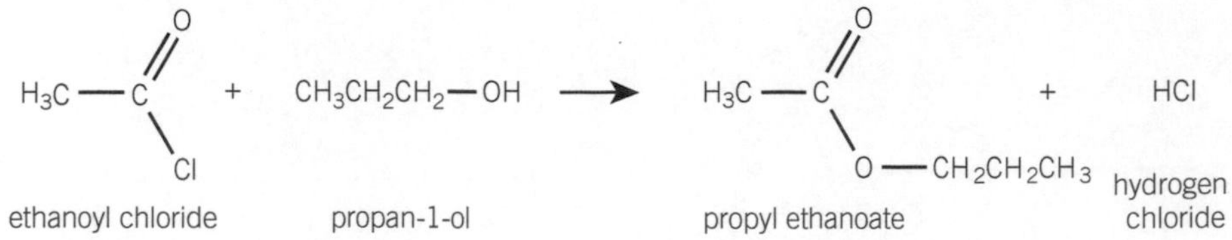

▲ **Figure 8** *The reaction of an acyl chloride with an alcohol to form an ester*

Acyl chlorides in synthesis

Acyl chlorides are used in the synthesis of carboxylic acid derivatives, e.g.:

- aliphatic and aromatic esters
- carboxylic acids
- primary and secondary amides.

The examples below show equations with ethanoyl chloride, CH_3COCl. But any acyl chloride would react in a similar way.

Acyl chlorides → esters

CH_3COCl	+	CH_3CH_2OH	→	$CH_3COOCH_2CH_3$	+	HCl
acyl chloride		**alcohol**		**aliphatic ester**		
CH_3COCl	+	C_6H_5OH	→	$CH_3COOC_6H_5$	+	HCl
acyl chloride		**phenol**		**aromatic ester**		

Acyl chlorides → carboxylic acids

CH_3COCl	+	H_2O	→	CH_3COOH	+	HCl
acyl chloride		**water**		**carboxylic acid**		

Acyl chlorides → amides

CH_3COCl	+	$2NH_3$	→	CH_3CONH_2	+	$NH_4^+Cl^-$
acyl chloride		**ammonia**		**primary amide**		
CH_3COCl	+	$2CH_3NH_2$	→	$CH_3CONHCH_3$	+	$CH_3NH_3^+Cl^-$
acyl chloride		**amine**		**secondary amide**		

Reactions of acid anhydrides

Acid anhydrides react with alcohols, phenols, water, ammonia, and amines in a similar way to acyl chlorides.

- Acid anhydrides are less reactive than acyl chlorides and are useful for laboratory preparations where acyl chlorides may be too reactive. The second product is the parent carboxylic acid of the acid anhydride.
- The equation for the formation of phenyl ethanoate from ethanoic anhydride and phenol is shown below.

$(CH_3CO)_2O$	+	C_6H_5OH	→	$CH_3COOC_6H_5$	+	CH_3COOH
ethanoic anhydride		phenol		phenyl ethanoate		ethanoic acid

Revision tip

Esters of phenols can be prepared from acyl chlorides or acid anhydrides. Phenols are not readily esterified by carboxylic acids.

Revision tip

You need to learn all the reactions of acyl chlorides shown here, including reagents and products. You should be able to write equations for reactions of acyl chlorides for each transformation.

- Reaction with alcohols, phenols, and water produces HCl as the second product.
- In the presence of a base such as ammonia and amines, any HCl initially produced forms an ammonium salt.

Synoptic link

For details of primary, secondary, and tertiary amides, see Topic 27.2, Amino acids, amides, and chirality.

Summary questions

1 Name the following esters.
 a $CH_3CH_2CH_2COOCH_3$ *(1 mark)* **b** $CH_3CH_2COO(CH_2)_4CH_3$ *(1 mark)* **c** $HCOOCH_2CH_2CH_3$ *(1 mark)*

2 **a** Write equations for the preparation of the following esters from a carboxylic acid and an alcohol.
 i propyl butanoate *(1 mark)* **ii** hexyl methanoate. *(1 mark)*
 b Write an equation for the formation of propanoyl chloride from a carboxylic acid. *(1 mark)*
 c Write equations for the hydrolysis of propyl propanoate by:
 i acid hydrolysis *(1 mark)* **ii** alkaline hydrolysis. *(1 mark)*

3 Write equations for the following preparations.
 a ethyl benzoate from an acyl chloride *(1 mark)*
 b pentyl propanoate from an acid anhydride *(1 mark)*
 c butanamide from a carboxylic acid (two steps). *(2 marks)*

Chapter 26 Practice questions

1 Which structure is ethyl 3-methylbutanoate?

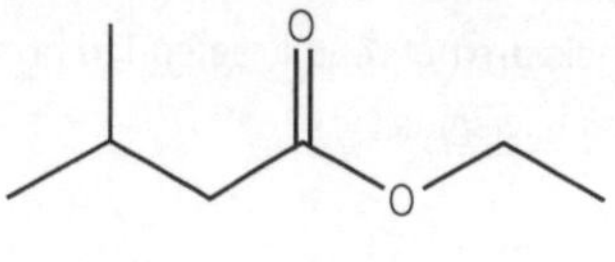

A

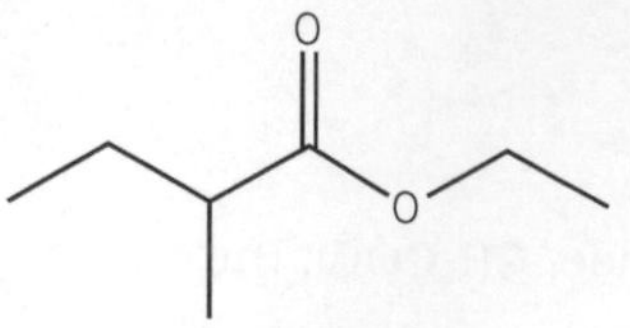

B

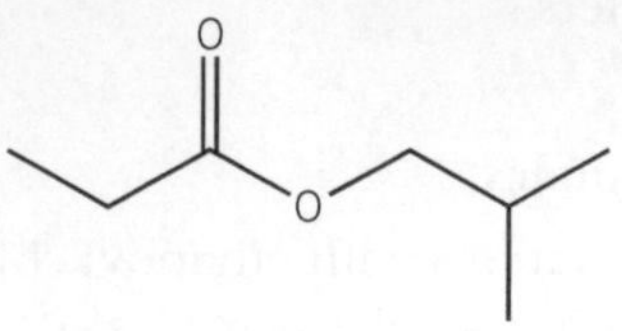

C

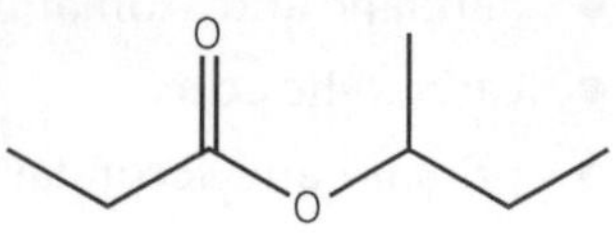

D *(1 mark)*

2 An organic compound **A** is oxidised by using acidified $K_2Cr_2O_7$. The organic product forms an orange precipitate with 2,4-DNP but does not produce a silver mirror with Tollens' reagent.

Which compound could be **A**?

A $(CH_3)_2CHOH$ **B** CH_3CH_2OH **C** CH_3CHO **D** $(CH_3)_2CHCOCH_3$ *(1 mark)*

3 Which reagent does **not** react with both aldehydes and ketones?

A $NaCN/H_2SO_4$ **B** 2,4-DNP **C** $AgNO_3/NH_3$ **D** $NaBH_4$ *(1 mark)*

4 Which reaction does **not** form propanoic acid as a product.

A CH_3CH_2CHO and $H^+(aq)/Cr_2O_7^{2-}(aq)$

B $CH_3CH_2COOCH_2CH_2CH_3$ and NaOH(aq)

C CH_3CH_2COCl and H_2O

D $CH_3CH_2(CO)O(CO)CH_2CH_3$ and CH_3OH *(1 mark)*

5 This question is about structures **E** and **F** in Figure 1.

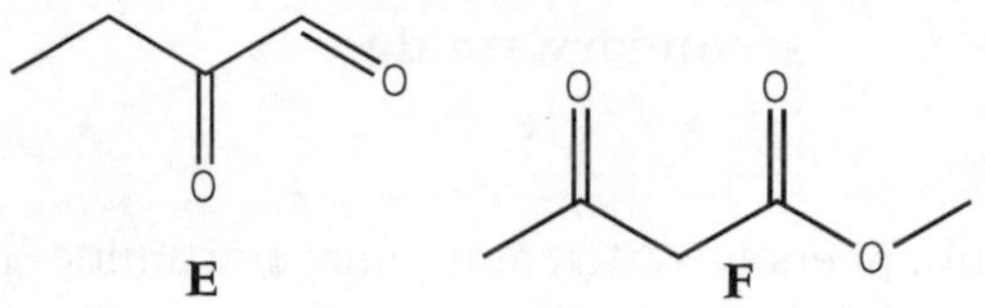

E **F**

▲ **Figure 1** *Structures E and F*

a For each reaction of **E**, state the reagents and equation.

i oxidation *(3 marks)* **ii** reduction *(3 marks)*

b Butanone is reacted with $NaCN/H_2SO_4$.

i Draw the structure of the organic product. *(1 mark)*

ii Name the functional groups in the organic product. *(2 marks)*

iii Name the reaction mechanism. *(1 mark)*

iv Outline the mechanism for this reaction. *(3 marks)*

c Write equations for the acid and alkaline hydrolysis of compound **F**. *(4 marks)*

6 a Write the equations for the preparation of the following esters from a carboxylic acid and an alcohol:

i pentyl propanoate *(2 marks)*

ii $CH_3COOCH(CH_3)CH_2CH_3$ *(2 marks)*

iii $C_6H_5COOCH_2C_6H_5$ *(2 marks)*

b Write equations for the following reactions of carboxylic acids and their derivatives.

i propanoic acid and calcium carbonate *(1 mark)*

ii ethanoyl anhydride and propan-2-ol *(2 marks)*

iii ethanoyl chloride and ammonia. *(2 marks)*

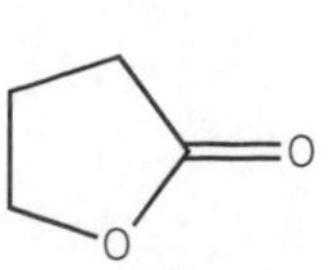

G

c Compound **G** can be prepared by esterification of a single organic compound.

State the reagent(s) and the formula of the organic compound. *(2 marks)*

27.1 Amines

Specification reference: 6.2.1

Amines

Amines are organic bases containing C, H, and N. Amines are derived from ammonia, NH_3. In an amine, one or more hydrogen atoms in NH_3 have been replaced by a carbon chain or aromatic ring.

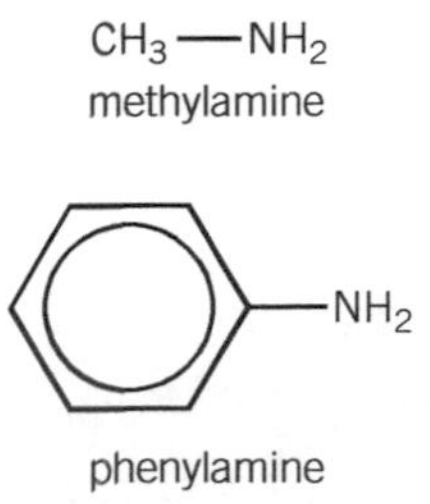

▲ **Figure 1** *Methylamine (top), an aliphatic primary amine; phenylamine (bottom), an aromatic primary amine*

Aliphatic amines

- A **primary** amine, RNH_2, has **one** hydrogen atom in NH_3 replaced by **one** alkyl chain, e.g. CH_3NH_2 (methylamine).
- A **secondary** amine, R_2NH, has **two** hydrogen atoms in NH_3 replaced by **two** alkyl chains, e.g. $(CH_3)_2NH$ (dimethylamine).
- A **tertiary** amine, R_3N, has all **three** hydrogen atoms in NH_3 replaced by **three** alkyl chains, e.g. $(CH_3)_3N$ (trimethylamine).

Aromatic amines

- In an **aromatic** amine, the NH_2 group is bonded to a benzene ring, e.g. $C_6H_5NH_2$ (phenylamine).

Amines as bases

Amines behave as bases.

- The lone pair of electrons on the N atom can accept a proton.
- A dative covalent bond forms between the lone pair on the N atom and H^+ (Figure 2).

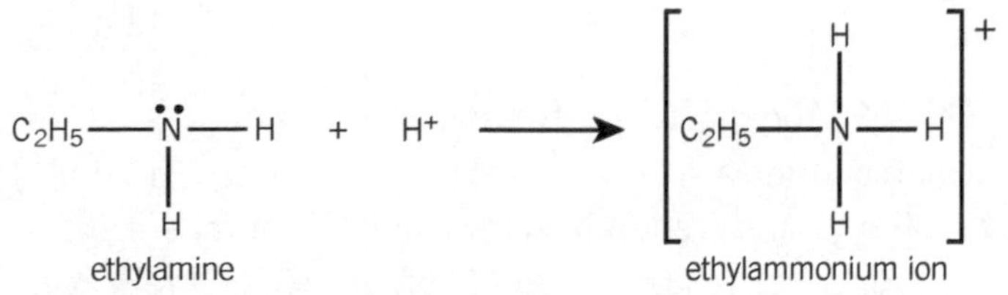

▲ **Figure 2** *Reaction of ethylamine as a base*

Neutralisation

As bases, amines neutralise dilute acids to form salts. The reaction of ethylamine with hydrochloric acid forms the salt, ethylammonium chloride:

$$CH_3CH_2NH_2 + HCl \rightarrow CH_3CH_2NH_3^+Cl^-$$

ethylammonium chloride

Synoptic link

The neutralisation of acids by amines is like the neutralisation of acids by ammonia:

$NH_3 + HCl \rightarrow NH_4^+Cl^-$

For details of neutralisation with ammonia, see Topic 4.1, Acids, bases, and neutralisation.

Preparation of amines

Aliphatic amines

Aliphatic primary amines are prepared by:

- substitution of haloalkanes with excess ethanolic ammonia to form a salt:

$$CH_3CH_2Cl + NH_3 \rightarrow CH_3CH_2NH_3^+Cl^-$$

salt

- addition of aqueous alkali to the mixture to generate the amine:

$$CH_3CH_2NH_3^+Cl^- + NaOH \rightarrow CH_3CH_2NH_2 + NaCl + H_2O$$

primary amine

Ethanol is a solvent, used to prevent substitution of the haloalkane by water to produce alcohols.

Synoptic link

See Topic 15.1, The chemistry of haloalkanes, for details of nucleophilic substitution reactions of haloalkanes.

Revision tip

Excess ammonia prevents further substitution of the amine group to form secondary and tertiary amines.

Aromatic amines

Aromatic amines are prepared by:

- reduction of nitroarenes by refluxing with tin and concentrated hydrochloric acid to form a salt.
- addition of aqueous alkali to the mixture to generate the amine.

The reduction of nitrobenzene is shown in Figure 3.

Revision tip

In the equation, the reducing agent (Sn/conc. HCl) is shown as [H].

Take care with balancing this equation.

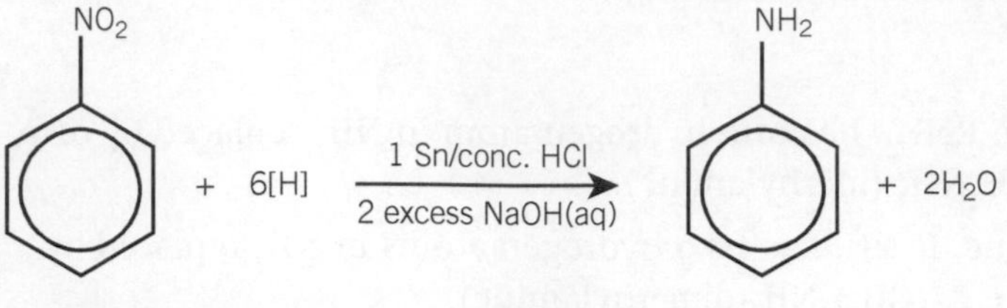

▲ **Figure 3** *The reduction of nitrobenzene to form phenylamine*

Summary questions

1 a Explain how amines behave as bases. *(2 marks)*

b Phenylamine is added to hydrochloric acid.

i Write the equation. *(1 mark)*

ii Name the salt formed. *(1 mark)*

2 Butylamine can be prepared from a haloalkane by two steps.

a Name the starting haloalkane and state the reagents for both steps. *(3 marks)*

b Write the formula of the organic product formed after the first step. *(1 mark)*

3 a Write an equation for the reaction between an excess of phenylamine and sulfuric acid. *(1 mark)*

b An aromatic amine can be prepared by refluxing 1,3-nitrobenzene with excess Sn/conc. HCl, followed by addition of excess NaOH(aq). Write an overall equation for this reaction. *(2 marks)*

27.2 Amino acids, amides, and chirality

Specification reference: 6.2.2

Amino acids

An amino acid is an organic compound containing C, H, N, and O. Amino acids contain both amine, NH_2, and carboxylic acid, COOH, functional groups.

The body has 20 common α-amino acids that can be built into proteins.

- An α-amino acid has the NH_2 and COOH groups attached to the same α-carbon atom (Figure 1).
- Different α-amino acids have different side chains, R, attached to the same α-carbon atom.

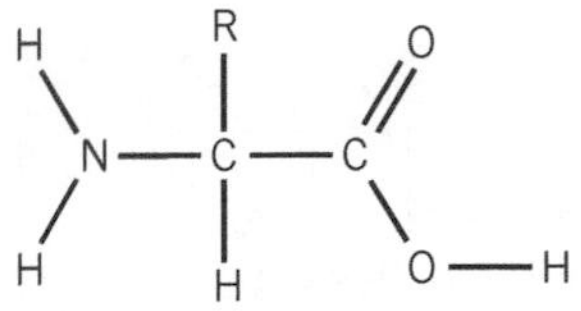

▲ **Figure 1** *The structure of an α-amino acid*
The α-carbon is the carbon atom next to the COOH group.

General formula

The general formula of an α-amino acid can be written as $RCH(NH_2)COOH$.

Amino acids have both an acidic COOH and a basic NH_2 functional group. So, amino acids have similar reactions to both carboxylic acids and amines.

Reactions of the COOH group in amino acids

Reaction with alkalis

The COOH group in an amino acid reacts with an aqueous alkali to form a salt and water:

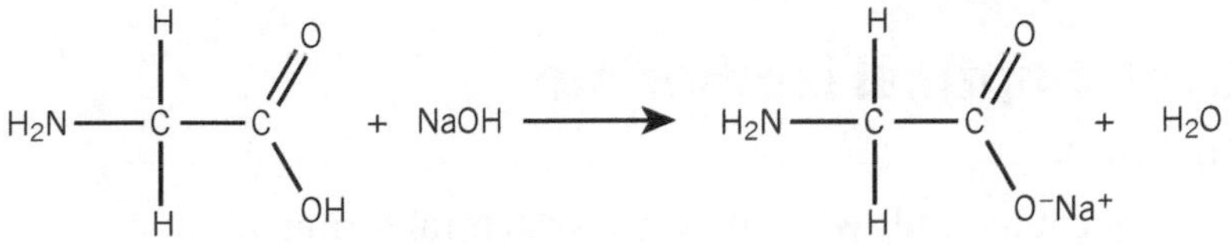

▲ **Figure 2** *The reaction of glycine (R = H) with aqueous sodium hydroxide*

Synoptic link

For details of formation of carboxylate salts, See Topic 26.3, Carboxylic acids.

Esterification with alcohols

The carboxylic acid group in amino acids can be esterified by heating with an alcohol in the presence of a concentrated sulfuric acid catalyst.

In Figure 3, the α-amino acid serine (R = CH_2OH) is reacted with excess ethanol and an acid catalyst. Notice that the acidic conditions then protonate the basic amine group to form $–NH_3^+$.

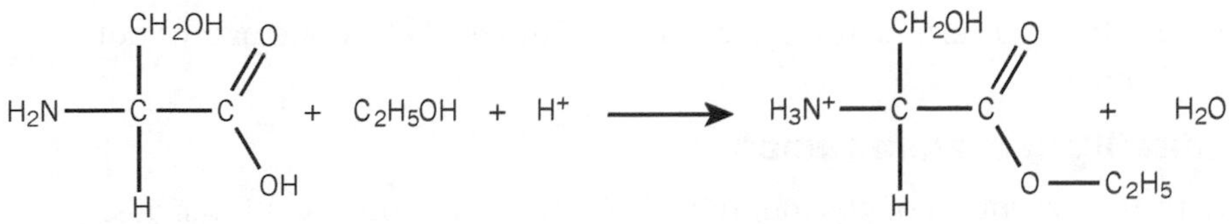

▲ **Figure 3** *Serine (R = CH_2OH) reacts with ethanol to form an ester – esterification*

Synoptic link

For details of esterification of carboxylic acids, see Topic 26.4, Carboxylic acid derivatives.

Reactions of the NH_2 group in amino acids

Reaction with acids

The NH_2 group in an amino acid neutralises an acid to form a solution of a salt:

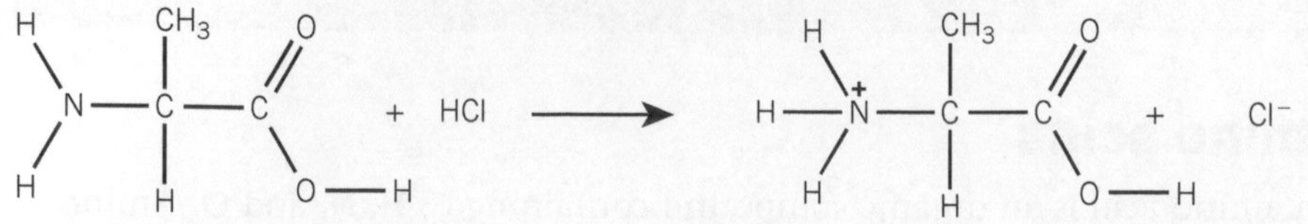

▲ **Figure 4** *The reaction of alanine (R = CH_3) with hydrochloric acid*

> **Synoptic link**
>
> For details of formation of amine salts, See Topic 27.1, Amines.

Amides

Amides are organic compounds containing C, H, N, and O that are derivatives of carboxylic acids. As with amines, there are primary, secondary, and tertiary amides (see Figure 5).

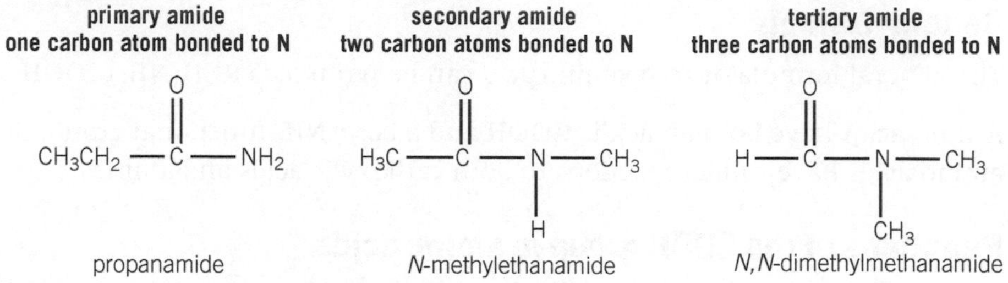

▲ **Figure 5** *Primary, secondary, and tertiary amides*

Amides are very stable compounds, found naturally in proteins. Synthetic amides are used as polyamides for clothing.

> **Synoptic link**
>
> Amides can be prepared by reacting acyl chlorides with ammonia and amines. See Topic 26.4, Carboxylic acid derivatives.
>
> For details of polyamides, see Topic 27.3, Condensation polymers.

Chirality and optical isomerism

Stereoisomerism

Stereoisomers are compounds with the same structural formula but a different arrangement of atoms in space. There are two types of stereoisomerism: optical isomerism and *E*/*Z* isomerism.

> **Synoptic link**
>
> See Topic 13.2, Stereoisomerism, for an introduction to stereoisomerism and *E*/*Z* isomerism.
>
> You have also come across optical isomers in transition elements: see Topic 24.3, Stereoisomerism in complex ions.

Optical isomerism

In organic chemistry, optical isomerism is found in molecules that contain a carbon atom that is a chiral centre.

- A chiral carbon atom is attached to four different atoms or groups of atoms.
- The four groups attached to the chiral carbon are arranged in space as two non-superimposable mirror images called **optical isomers**.
- Each chiral carbon atom in an organic molecule has one pair of optical isomers.

Chirality in α-amino acids

With the exception of glycine, H_2NCH_2COOH, all of the α-amino acids, $RCH(NH_2)COOH$, contain a chiral carbon atom.

- The α-carbon atom is bonded to four different atoms or groups of atoms, NH_2, H, COOH, and R.
- In diagrams, chiral carbon atoms are often labelled with an asterisk * (see Figure 6).

> **Revision tip**
>
> Like a pair of hands, optical isomers can be considered as right- and left-handed forms.
>
> One optical isomer cannot be superimposed upon the other.

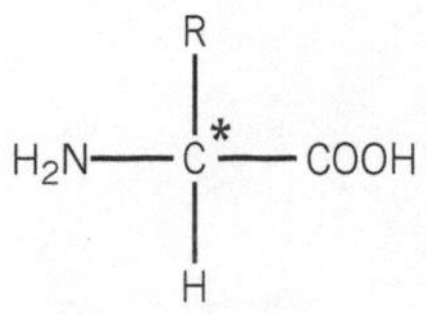

▲ **Figure 6** *The chiral carbon atom in an α-amino acid*

Drawing optical isomers

Optical isomers are drawn showing the 3D tetrahedral arrangement of the four different groups around the central chiral carbon atom.

- First draw one optical isomer, taking care to show bonds to the bonded atoms.
- Now draw the other optical isomer as a mirror image, reflecting the first structure.

The two optical isomers of the amino acid alanine, $CH_3CH(NH_2)COOH$, are shown in Figure 7.

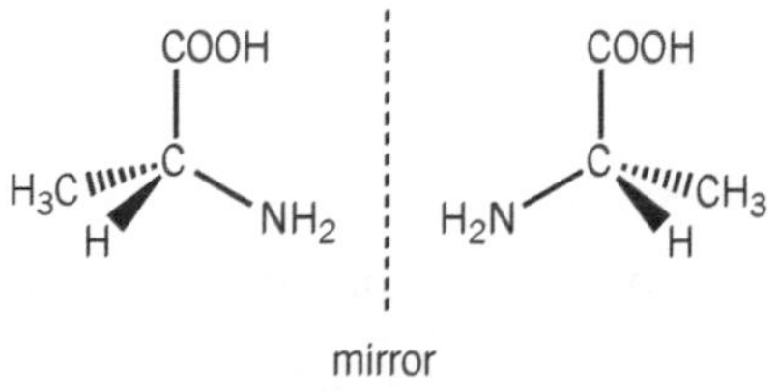

▲ **Figure 7** *Optical isomers of the amino acid alanine ($R = CH_3$)*
The diagrams show the 3D arrangement of the four different groups around a chiral carbon atom

Chirality in other organic compounds

Chiral carbon atoms exist widely in naturally occurring organic molecules and are not restricted to just α-amino acids. For example, all sugars, proteins, and nucleic acids contain chiral carbon atoms.

You should be able to identify chiral centres in a molecule of any organic compound by identifying the carbon atoms that are connected to four different atoms or groups.

Summary questions

1 a How many chiral carbons are in the following?
 i $CH_3CH(OH)CH_2CH_3$ *(1 mark)*
 ii $CH_3CH(OH)CH(OH)CH_2CH_3$ *(1 mark)*
b Why does glycine (R = –H), not have optical isomers? *(1 mark)*

2 a Write the structural formula of the α-amino acids with the following R groups.
 i $-CH_2SH$ *(1 mark)* ii $-CH(CH_3)_2$ *(1 mark)*
b What is the organic product of the following reactions of the amino acid alanine ($R = -CH_3$)?
 i With NaOH(aq). *(1 mark)*
 ii With CH_3OH and an acid catalyst. *(1 mark)*

3 a Draw the 3D structures for the optical isomers in the following.
 i 2-bromobutane *(2 marks)*
 ii 2-hydroxybutanoic acid *(2 marks)*
b The amino acid lysine has the R group $-(CH_2)_4NH_2$. Lysine is heated with methanol and an acid catalyst to form an organic product **A**. Draw the structure of:
 i lysine *(1 mark)* ii the organic product **A**. *(2 marks)*

Synoptic link

See Topic 6.1, Shapes of molecules and ions, for details of drawing 3D shapes using wedges.

Revision tip

Wedges are essential to display the 3D diagrams accurately.

Take care that all bonds connect to the correct atoms.

Revision tip

Many organic compounds have more than one chiral carbon atom.

Number of optical isomers = 2^n

n = number of chiral carbon atoms

$n = 1$ gives $2^1 = 2$

$n = 2$ gives $2^2 = 4$

$n = 3$ gives $2^3 = 8$

and so on.

27.3 Condensation polymers

Specification reference: 6.2.3

Synoptic link

You covered addition polymerisation in Topic 13.5, Polymerisation in alkenes.

See Topic 26.4, Carboxylic acid derivatives, for details of esterification.

Addition and condensation polymerisation

Addition polymerisation is the formation of a very long molecular chain by repeated addition reactions of many unsaturated alkene molecules (monomers).

Condensation polymerisation is the joining of monomers with loss of a small molecule, usually water (the condensation) or hydrogen chloride.

Polyesters and polyamides are two important condensation polymers that are derivatives of carboxylic acids.

Revision tip

Polyesters and polyamides can also be prepared from acyl chlorides instead of carboxylic acids. The process is essentially the same with an HCl molecule, rather than a H_2O molecule, being formed for each ester or amide linkage.

Acyl chlorides have the advantage of being more reactive than carboxylic acids and giving a higher yield.

Polyesters

In a polyester, the monomers have been joined together with ester linkages to form the polymer.

Polyesters can be made from:

- one monomer containing both a carboxylic acid and a hydroxyl group, or
- two monomers, one containing two carboxylic acid groups and the other containing two hydroxyl groups.

Polyesters from one monomer

Polyesters can be made from **one** monomer containing **both** a carboxylic acid and a hydroxyl group.

Glycolic acid, $HOCH_2COOH$, contains both –COOH and –OH groups.

- The –COOH group in one molecule of glycolic acid reacts with the –OH group of another molecule of glycolic acid.
- An ester linkage forms between the two molecules, together with water.
- This is repeated many thousands of times to form the polymer (Figure 1).

Revision tip

You are **not** expected to recall the structures of actual polyesters and polyamides or their monomers. But you are required to apply the principles of condensation polymerisation.

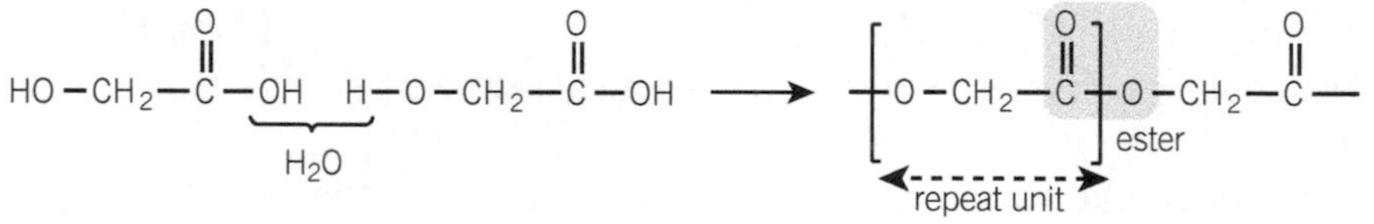

▲ **Figure 1** *Condensation polymerisation of glycolic acid, $HOCH_2COOH$, showing two repeat units*

Revision tip

Formation of a polyester is essentially esterification repeated on a giant scale.

The formation of water for each linkage given condensation polymerisation its name.

Polyesters from two monomers

Polyesters can be made from two different monomers:

- one monomer containing two carboxylic acid groups (a dicarboxylic acid)
- the other monomer containing two hydroxyl groups (a diol).
- One of the –COOH groups in a molecule of the dicarboxylic acid reacts with one of the –OH groups in a molecule of the diol.
- This process is repeated many thousands of times to form the polymer.

Figure 2 shows the formation of a polyester from its two monomers, hexanedioic acid (the dicarboxylic acid) and hexane-1,6-diol (the diol).

Revision tip

The formation of a polyester builds a polymer from –COOH in one molecule and –OH in another molecule.

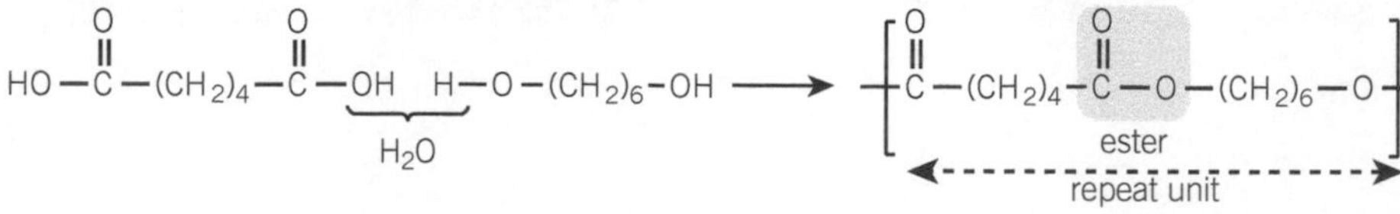

▲ **Figure 2** *The formation of a polyester from two monomers showing one repeat unit*

Polyamides

In a polyamide, the monomers have been joined together with amide linkages to form the polymer.

Polyamides can be made from:

- one monomer containing both a carboxylic acid and an amine group, or
- two monomers, one containing two carboxylic acid groups and the other containing two amine groups.

Revision tip

Polyamide formation is essentially the same process as formation of polyesters. The only real difference is the $-NH_2$ group for polyamides and the $-OH$ group for polyesters.

Polyamides from amino acids

Amino acids contain **both** an amine group and a carboxylic acid group. Amino acids undergo condensation polymerisation to form polypeptides or proteins containing many different amino acids all linked together by amide bonds.

An amino acid, $RCH(NH_2)COOH$, contains both $-COOH$ and $-NH_2$ groups.

- The $-COOH$ group in one amino acid molecule reacts with the $-NH_2$ group of another molecule of an amino acid.
- An amide linkage forms between the two molecules, together with water.
- This is repeated many thousands of times to form the polymer (Figure 3).

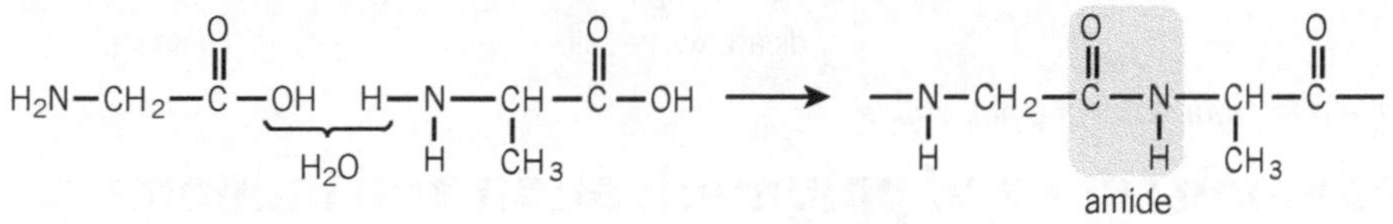

▲ **Figure 3** *The formation of a section of a protein from two different amino acids (R = H and R = CH_3)*

Polyamides from two monomers

As with polyesters, polyamides can be made from two different monomers:

- One of the $-COOH$ groups in one molecule of the dicarboxylic acid reacts with one of the $-NH_2$ groups in a molecule of the diamine.
- This process is repeated many thousands of times to form the polymer.

Figure 4 shows the formation of Nylon 6,6 from its two monomers, hexanedioic acid (the dicarboxylic acid) and 1,6-diaminohexane (the diamine).

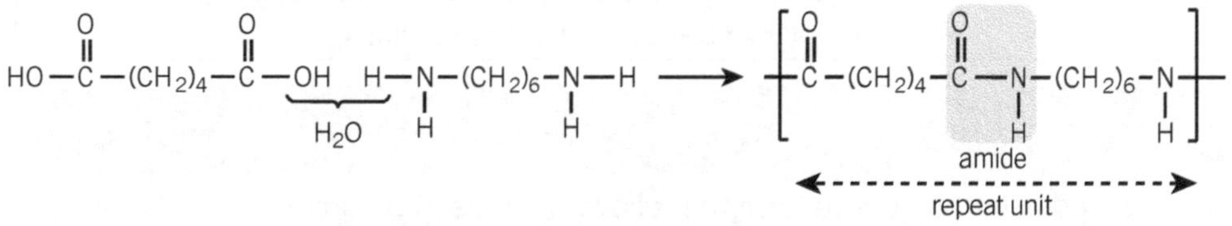

▲ **Figure 4** *Synthesis of Nylon 6,6 from the reaction of a diamine with a dicarboxylic acid*

Hydrolysis of condensation polymers

Polyesters and polyamides can be broken down by hydrolysis using:

- hot aqueous acid such as hydrochloric acid (acid hydrolysis), or
- hot aqueous alkali such as sodium hydroxide (alkaline hydrolysis).

Synoptic link

This is the same principle as for acid and alkaline hydrolysis of esters. For details, see Topic 26.4, Carboxylic acid derivatives.

Hydrolysing polyesters

The acid and the base hydrolysis of a polyester is shown below.

- Acid hydrolysis produces a carboxylic acid and an alcohol.
- Base hydrolysis produces a carboxylate salt and an alcohol.

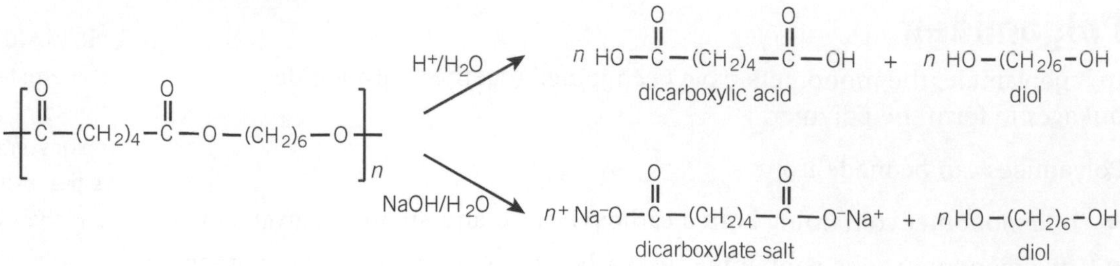

▲ **Figure 5** *The acid and base hydrolysis of a polyester*

Hydrolysing polyamides

The acid and base hydrolysis of a polyamide is shown below.

- Acid hydrolysis produces a carboxylic acid and an ammonium salt.
- Base hydrolysis produces a carboxylate salt and an amine.

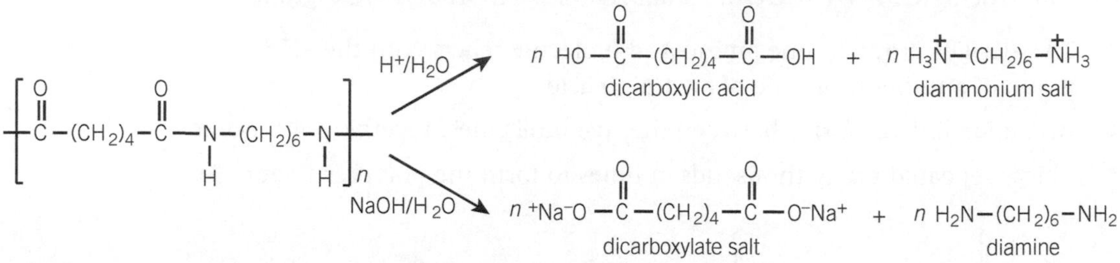

▲ **Figure 6** *The acid and base hydrolysis of a polyamide*

Revision tip

Take great care with the products of acid and base hydrolysis:

- base hydrolysis of polyesters **and** polyamides produces carboxylate salts
- acid hydrolysis of polyamides produces ammonium salts.

Predicting types of polymerisation and monomers

You should be able to:

- predict the type of polymerisation taking place given the monomer(s), and
- identify monomers from polymer chains.

Table 1 outlines what to look for when deciding the type of polymerisation.

▼ **Table 1** *Characteristics of addition and condensation polymerisation*

Type of polymerisation	Characteristics
Addition	Monomer contains a C=C double bond. Main polymer chain is a continuous chain of carbon atoms.
Condensation	Two monomers, each with two functional groups. One monomer with two different functional groups. Polymer contains ester or amide linkages.

Summary questions

1 a State the two functional groups needed in monomers to form the following types of polymer.
i polyamide **ii** polyester *(2 marks)*
b What is meant by **i** addition polymerisation **ii** condensation polymerisation? *(2 marks)*

2 a Draw structures to show one repeat unit of a polymer from the following monomers.
i $H_2NCH(CH_3)COOH$ **ii** $HOCH(C_6H_5)COOH$ *(4 marks)*
iii $HO(CH_2)_3OH$ and $HOOCCH(C_2H_5)COOH$ **iv** $H_2NCH(CH_3)NH_2$ and $HOOC(CH_2)_4COOH$ *(4 marks)*

3 Repeat units of two polymers are shown below.

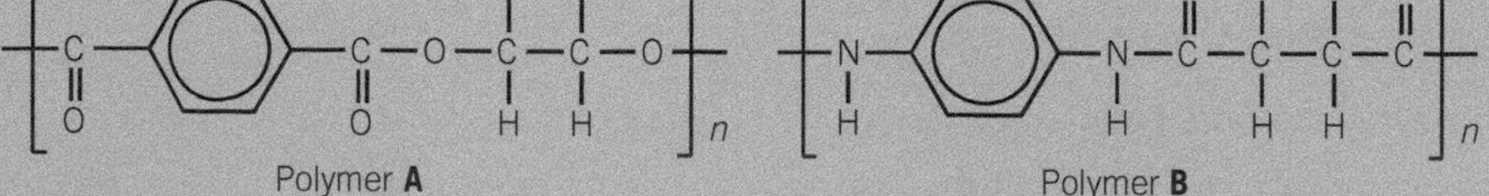

Polymer **A** Polymer **B**

a What are the products of base hydrolysis of the polymer **A**? *(2 marks)*
b What are the products of acid hydrolysis of the polymer **B**? *(2 marks)*

Chapter 27 Practice questions

1 What is the structure of aspartic acid (R = $-CH_2COOH$) at high pH?

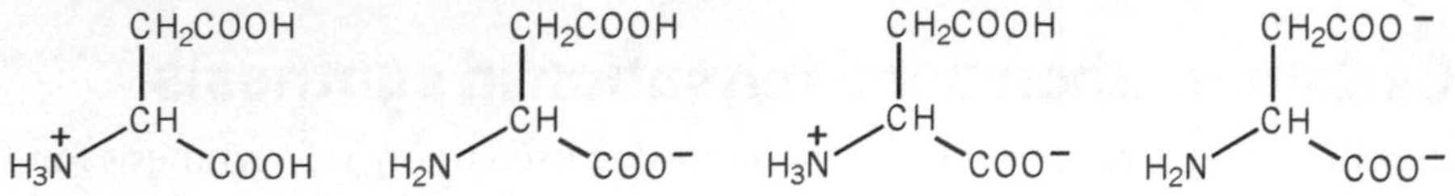

A **B** **C** **D**

(1 mark)

2 What is the number of optical isomers of compound **E**?

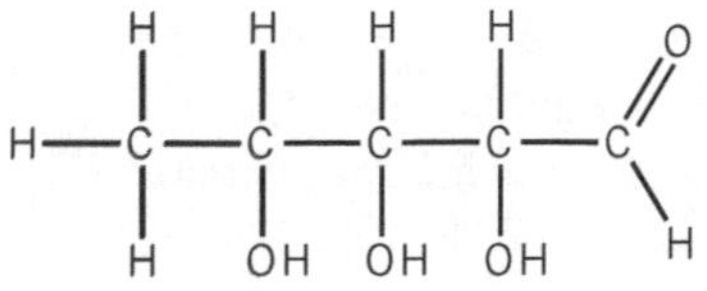

E

A 2 **B** 4 **C** 6 **D** 8 *(1 mark)*

3 The repeat unit of a polymer **F** is shown.

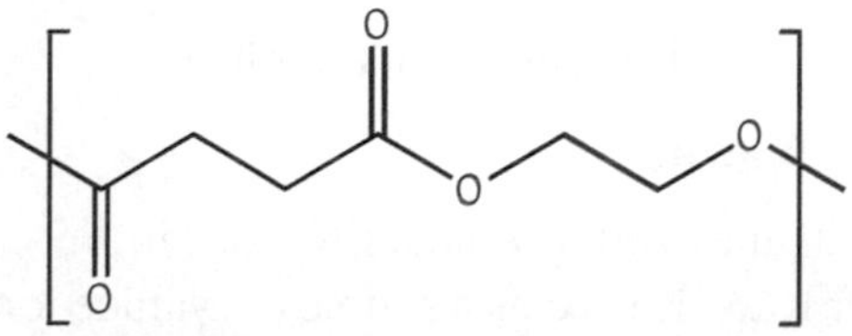

What are the monomers?

A ethane-1,2-diol and butanedioic acid

B butane-1,2-diol and ethanedioic acid

C 2-hydroxyethanoic acid

D 2-hydroxybutanoic acid *(1 mark)*

4 This question is about compounds with the amine functional group.

a **i** Explain how amines can act as bases. *(2 marks)*

ii Write an equation for the reaction of excess $C_6H_5NH_2$ with $H_2SO_4(aq)$. *(2 marks)*

b Propylamine can be prepared from a haloalkane.

i State the reagents and essential conditions. *(1 mark)*

ii Write an equation for the reaction. *(2 marks)*

c Phenylamine can be prepared by reduction.

i State the reagents and conditions for the reduction. *(1 mark)*

ii Write an equation for this formation of phenylamine. *(2 marks)*

d Polyamide **G** is hydrolysed.

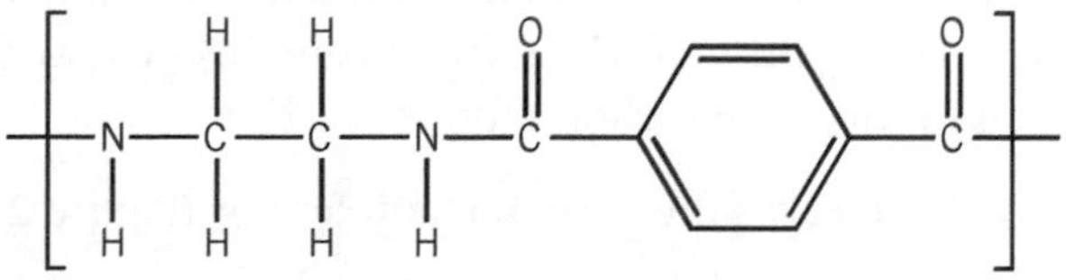

Draw the structures of the products of:

i base hydrolysis *(2 marks)*

ii acid hydrolysis. *(2 marks)*

5 This question is about amino acids.

a **i** Explain the term **optical isomers**. *(1 mark)*

ii Draw 3D diagrams for the optical isomers of serine (R = $-CH_2OH$). *(2 marks)*

b Draw the structure of the organic compound formed when alanine (R = $-CH_3$) reacts with methanol and an acid catalyst. *(2 marks)*

c A polymer has alternating molecules of the amino acids alanine (R = $-CH_3$) and serine (R = $-CH_2OH$).
Draw the repeat unit of this polymer. *(2 marks)*

28.1 Carbon–carbon bond formation

Specification reference: 6.2.4

Carbon–carbon bond formation in synthesis

Reactions that form carbon–carbon bonds are used in organic synthesis for:

- increasing the length of a carbon chain
- adding a carbon-containing side chain to a carbon chain or aromatic ring.

Nitriles

The nitrile group has the functional group –C≡N, commonly shown as –CN.

- Nitriles are named from the total number of carbon atoms in the chain.
- CH_3CH_2CN has a three-carbon chain and is named propanenitrile.

Preparation of nitriles from haloalkanes

Nitriles can be formed by reacting haloalkanes with cyanide, CN^-, ions in ethanol. An ionic cyanide, e.g. NaCN or KCN, is used as a source of cyanide ions.

For primary haloalkanes, with the halogen at the end of the carbon chain, the reaction **increases** the carbon chain length, e.g. C–Br → C–C≡N.

$CH_3CH_2CH_2Br + KCN \rightarrow CH_3CH_2CH_2CN + KBr$
1-bromopropane → butanenitrile
3 carbon atoms → **4** carbon atoms

Mechanism

The reaction mechanism is **nucleophilic substitution** (Figure 1):

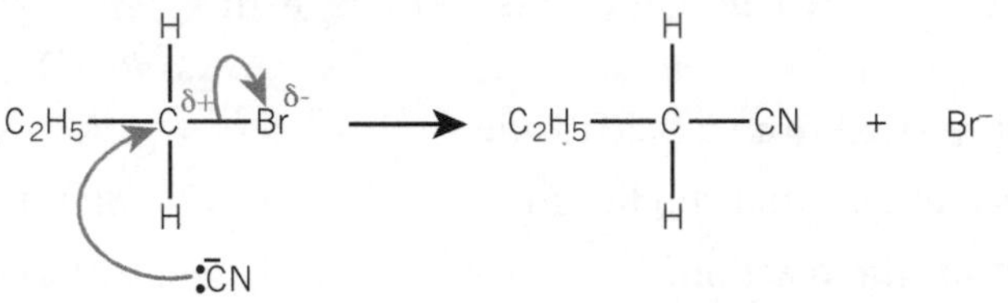

▲ **Figure 1** *Formation of a nitrile by nucleophilic substitution of a haloalkane*

Preparation of hydroxynitriles from carbonyl compounds

Hydrogen cyanide, HCN adds across the C=O bond of aldehydes and ketones to form a hydroxynitrile. HCN is generated in solution using sodium cyanide, NaCN and sulfuric acid, H_2SO_4. The hydroxynitrile formed contains –OH and –CN functional groups bonded to the same carbon atom.

Addition of HCN to an aldehyde increases the carbon chain length (Figure 2):

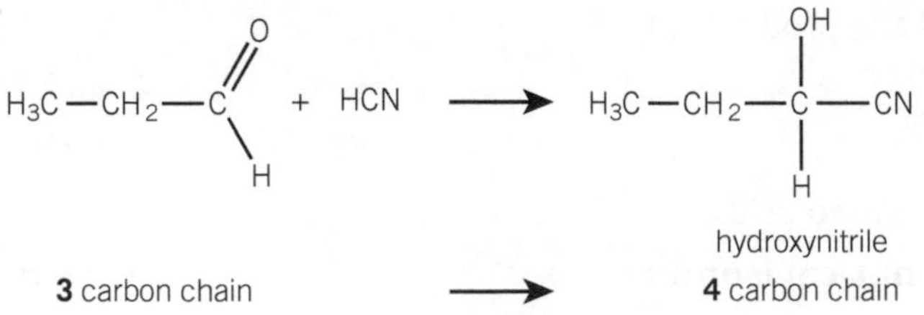

▲ **Figure 2** *Addition of HCN to an aldehyde increasing the carbon chain length*

Revision tip
When naming nitriles, the C atom in the C≡N group is counted as carbon-1 and is included within the longest carbon chain (similar to naming of carboxylic acids).

Revision tip
Ethanol is used as a solvent. If water is present, hydrolysis may take place to form an alcohol.

Revision tip
For secondary and tertiary haloalkanes, substitution introduces a carbon-containing side chain.

Synoptic link
You should recall the mechanism of nucleophilic substitution from Topic 15.1, The chemistry of the haloalkanes.

Revision tip
Take care with mechanisms containing cyanide ions, CN^-.

- Although the formula is normally written as CN^-, the lone pair and charge are on the C atom and not the N atom.
- Look closely at the CN^- ion and curly arrow in the mechanism.

Synoptic link
The reactions of aldehydes and ketones with hydrogen cyanide were discussed in Topic 26.1, Carbonyl compounds.

Addition of HCN to a ketone introduces a carbon-containing side chain to the carbon chain:

$$H_3C-\overset{\overset{\displaystyle O}{\|}}{C}-CH_3 \;+\; HCN \longrightarrow \; H_3C-\overset{\overset{\displaystyle OH}{|}}{\underset{\underset{\displaystyle CN}{|}}{C}}-CH_3$$

hydroxynitrile

3 carbon chain $\longrightarrow$ carbon side chain added

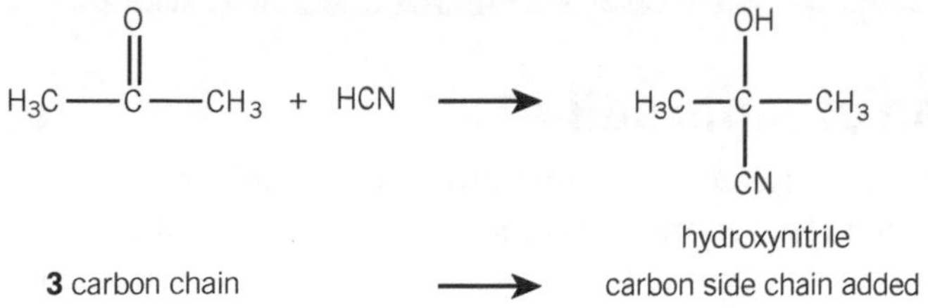

▲ **Figure 3** *Addition of HCN to a ketone adds a carbon containing side chain*

The reaction mechanism is **nucleophilic addition**.

Reactions of nitriles

The nitrile functional group can be easily converted into:

- amines by **reduction** $R–C{\equiv}N \rightarrow R–CH_2NH_2$
- carboxylic acids by **hydrolysis** $R–C{\equiv}N \rightarrow R–COOH$

Reduction of nitriles

Nitriles are reduced to amines with hydrogen in the presence of a nickel catalyst.

$$CH_3CH_2CH_2C{\equiv}N \;+\; 2H_2 \;\rightarrow\; CH_3CH_2CH_2CH_2NH_2$$
    butanenitrile   butylamine

Hydrolysis of nitriles

Nitriles are hydrolysed to carboxylic acids by heating with dilute aqueous acid, e.g. HCl(aq).

$$CH_3CH_2CH_2C{\equiv}N \;+\; 2H_2O \;+\; HCl \;\rightarrow\; CH_3CH_2CH_2COOH + NH_4Cl$$
    butanenitrile   butanoic acid

Forming carbon–carbon bonds to benzene rings

A carbon–carbon bond is formed to an aromatic ring (Ar) by

- alkylation, with a haloalkane, RBr $Ar \rightarrow Ar–R$
- acylation with an acyl chloride, RCOCl. $Ar \rightarrow Ar–COR$

These 'Friedel–Crafts' reactions require the presence of a halogen carrier.

> ### Synoptic link
> For details of the mechanism of nucleophilic addition of carbonyl compounds with HCN, see Topic 26.1, Carbonyl compounds.

> ### Revision tip
> Nitriles are useful intermediates in the synthesis of organic compounds.

> ### Revision tip
> Be careful when writing this equation. It is all too easy to include just one H_2.

> ### Synoptic link
> For details of these important reactions, see Topic 25.2, Electrophilic substitution reactions of benzene, where the alkylation and acylation of aromatic rings are discussed in detail.

Summary questions

1 Reactions of KCN in ethanol can increase carbon chain length or add a carbon-containing side chain.
   Use equations of KCN with isomers of C_3H_7Cl to illustrate this statement.
   (3 marks)

2 How could you prepare $CH_3CH_2CH(OH)COONa$ from a carbonyl compound (3 steps)?
   State the reagents, conditions, and equation.   *(6 marks)*

3 How could you synthesise the following from benzene?
   Include reagents, conditions, and equations.
   a $C_6H_5CH(CH_3)_2$ (1 step)   *(2 marks)*
   b C_6H_5COOH (2 steps)   *(4 marks)*
   c $C_6H_5CH(OH)CH_2CH_3$ (2 steps)   *(4 marks)*

28.2 Further practical techniques

Specification reference: 6.2.5

Synoptic link

Distillation and heating under reflux were covered in detail in Topic 16.1, Practical techniques in organic chemistry.

Revision tip

The solution may contain insoluble impurities (these will be visible). Solid impurities can be removed by filtering the hot solution quickly through fluted filter paper.

The solid impurities are trapped on the fluted filter paper.

Preparation of an organic solid

You have already seen how to use Quickfit apparatus for distilling and heating under reflux in the preparation of organic liquids. Organic solids are often prepared by refluxing a solution of the reactants. The organic product is obtained as an impure solid that needs to be purified.

Purification of an organic solid

The main purification steps are listed below.

- filtration under reduced pressure
- recrystallisation
- measurement of melting points

Filtration under reduced pressure

Filtration under reduced pressure is a technique for separating a solid product from a solvent or liquid reaction mixture. This technique uses a Buchner flask and funnel, filter paper, and access to a vacuum pump or filter pump connected to a water tap.

After the preparation, the reaction mixture is filtered under reduced pressure. The solid in the Buchner funnel is rinsed with cold solvent and partly dried under suction for a few minutes.

Recrystallisation

The filtered product will contain impurities, which are removed by recrystallisation.

The technique relies upon the following principles.

- The desired product is less soluble than soluble impurities in the chosen solvent.
- Solubility is greater in a hot solvent than a cold solvent.

The essential stages are listed below.

- A minimum volume of hot solvent is added to dissolve the impure solid.
- The resulting solution is cooled, allowing the product to crystallise out of solution.
- The crystals are filtered under reduced pressure and dried to obtain the pure solid.

Melting point determination

Chemists check the purity of a solid compound by measuring its melting point.

- A sample of the pure compound is placed in the bottom of a sealed capillary tube.
- The melting point is measured using an electric melting point apparatus or an oil-filled Thiele tube.

A pure organic substance usually has a very sharp melting point. Impure organic compounds have lower melting points, and melt over a wider temperature range, than the pure compound.

The melting point is compared with the value recorded in a database or table. If the melting point is lower than the data value, the sample is likely to be impure and would need to be recystallised again.

Summary questions

1 What apparatus would you use for filtration under reduced pressure? *(2 marks)*

2 Describe the difference in the melting points of impure and pure samples of a compound. *(2 marks)*

3 Describe the purpose of each stage in the purification of an impure solid from a reaction mixture. *(3 marks)*

Specification reference: 6.2.5

Identifying functional groups in molecules

It is essential that you can identify functional groups in molecules.

Figure 1 shows the aliphatic functional groups encountered in the A level course. Figure 2 shows some naturally occurring compounds with some of these functional groups.

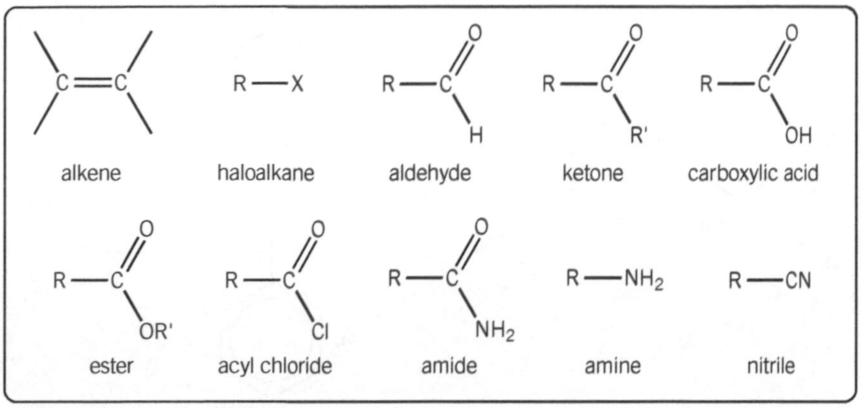

▲ **Figure 1** *Aliphatic functional groups*

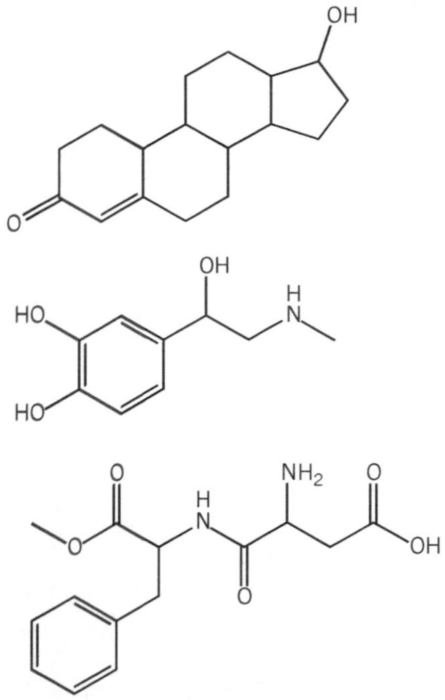

▲ **Figure 2** *Naturally occurring and synthetic organic compounds. Can you recognise the functional groups? (Answers with Summary answers)*

Predicting the reactions of organic molecules

As well as recognising functional groups, you need to learn all the reactions covered in the A level course. You should then be able to predict reactions and properties of any molecules containing these functional groups.

Reactions of aliphatic functional groups

Figure 3 shows the reactions of aliphatic functional groups.

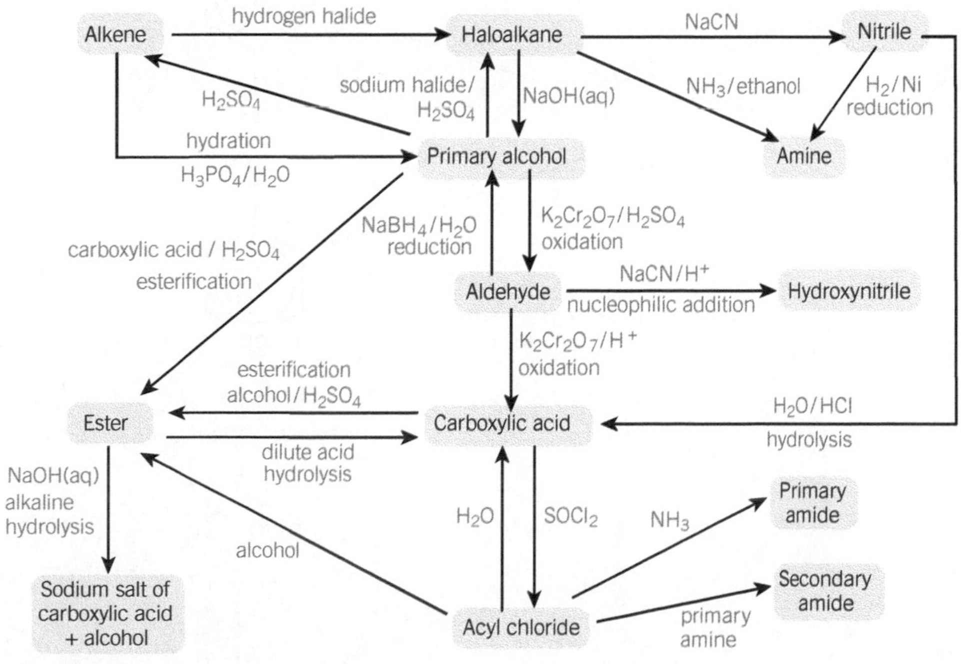

▲ **Figure 3** *Reactions of aliphatic functional groups*

Synoptic link

In Topic 16.2, Synthetic routes, you used flowcharts for the reactions in the first year of the course. It would be useful to review these and to link the reactions to the more extensive flowchart in Figure 3.

Revision tip

There are absolutely no shortcuts here.

- You need to learn the reagents and essential conditions for all these reactions. Try drawing out your own flowchart to see what you can remember (without cheating).
- You should also be able to write equations for the reactions. Try writing equations and then check from the details in the topics.
- If you don't know the functional group, learn it.

Reactions of aromatic functional groups

Figure 4 and Figure 5 show the reactions of aromatic functional groups.

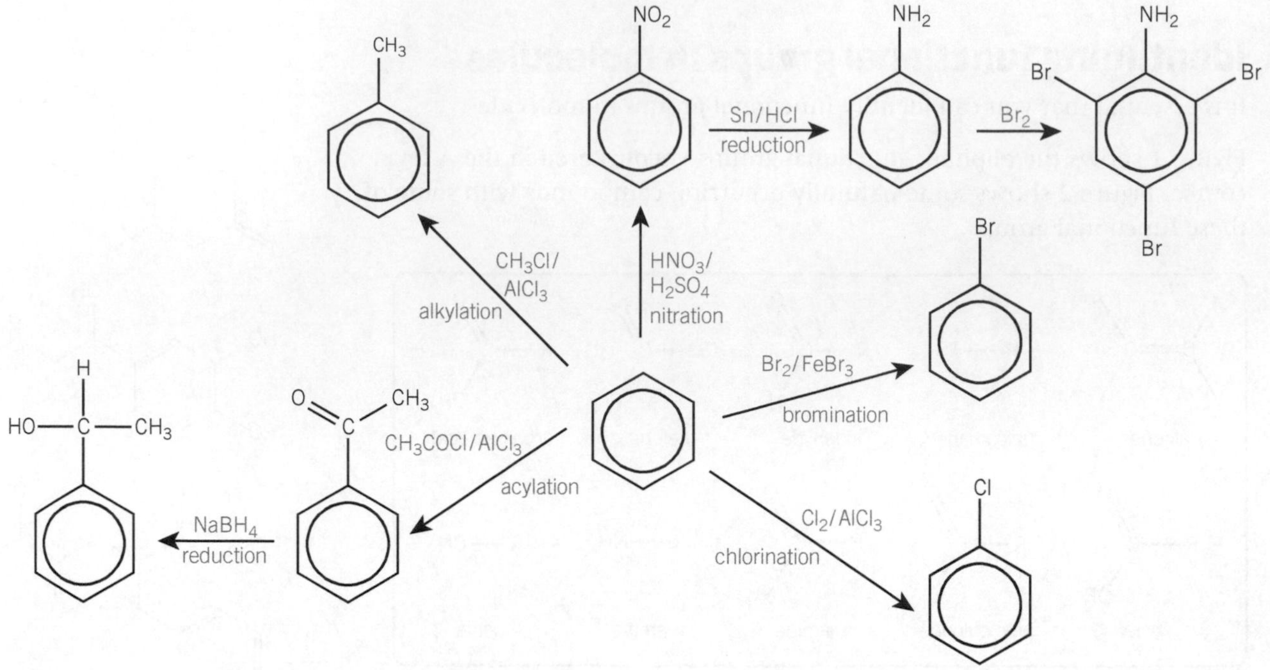

▲ **Figure 4** *Reactions of benzene and its compounds*

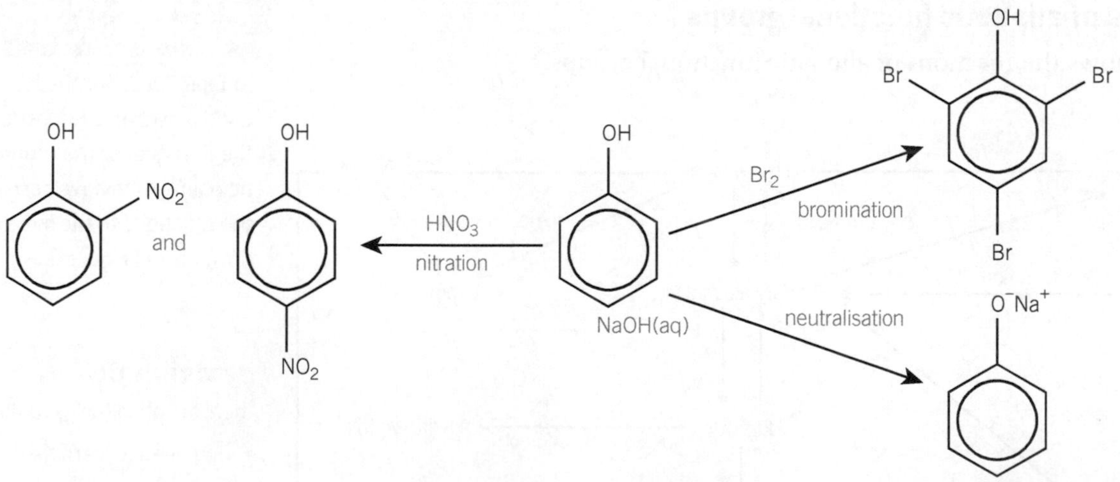

▲ **Figure 5** *Reactions of phenol*

Multi-stage synthetic routes

You are expected to be able to devise multi-stage synthetic routes for converting between all functional groups studied throughout the specification. In Topic 16.2, Synthetic routes, you devised two-stage synthetic routes using the functional groups encountered in the first year of the course. Now the list is far more extensive – there may be more stages but the principles are the same:

- Identify the functional groups in the starting and target molecules.
- Identify any intermediates that link the starting and target molecules. The intermediates needs to be linked by reagents and conditions that you have met during the course.

Worked example: Synthesis of propanenitrile

Devise a flowchart to show how a sample of propanenitrile could be prepared, starting from ethanal. Show the reagents you would use to carry out each stage in your flowchart.

Step 1: Identify the functional groups in the starting and target molecules.

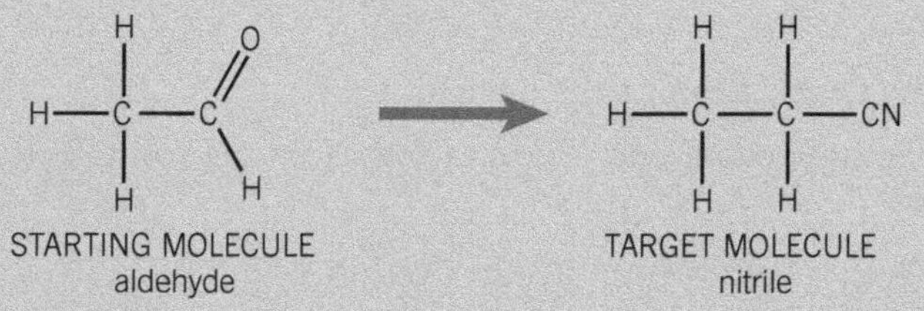

Step 2: Identify a sequence of chemical reactions that could convert the functional groups in the starting molecule into the functional groups of the target molecule.

- Nitriles can be prepared from haloalkanes.
- Haloalkanes can be prepared from alcohols.
- Alcohols can be prepared by reducing carbonyl compounds.

Step 3: The complete flowchart below shows the conversion of each functional group.

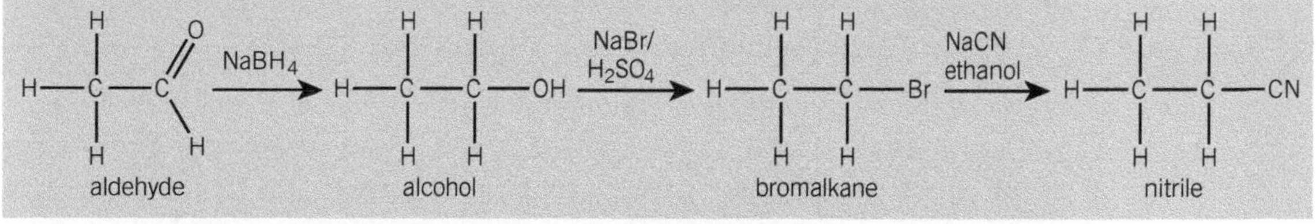

Revision tip

Look back at the flowchart in Figure 3 and find the aldehyde and nitrile functional groups.

You can then trace a route from aldehyde to nitrile by following the arrows for the reactions.

You can see the strength of the flowchart to help you learn.

Revision tip

The steps are easier to work out by working back from the target molecule.

Summary questions

1 Name the functional groups in compounds **A** and **B** below. *(2 marks)*

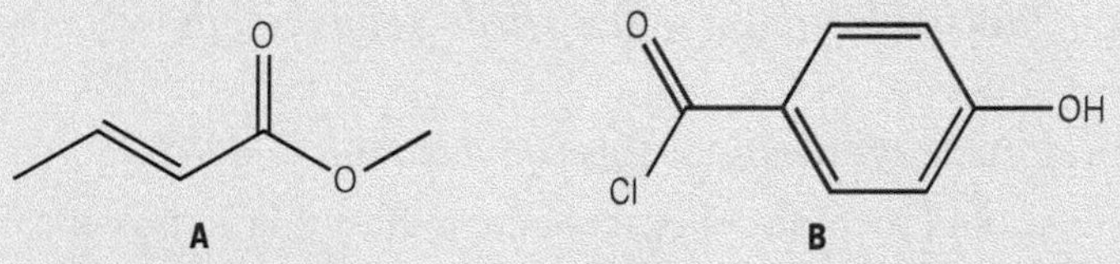

2 Plan a two-stage synthesis to prepare ethylamine, starting from ethene. For each stage, write an equation and state the reagents and conditions. *(4 marks)*

3 Plan a three-stage synthesis to prepare ethyl ethanoate, starting from bromoethane as the only organic compound. *(6 marks)*

Chapter 28 Practice questions

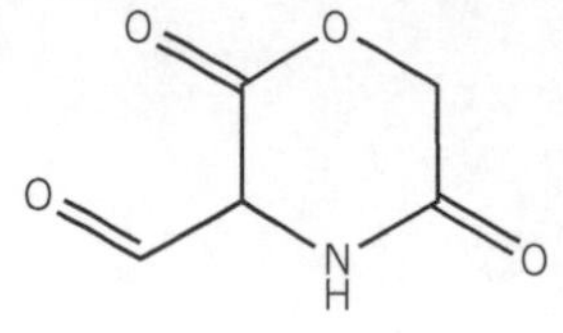

A

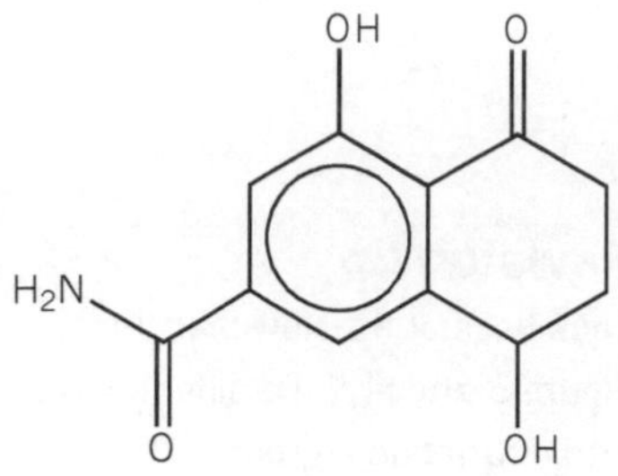

B

1 What are the functional groups in

a compound **A** *(3 marks)* **b** compound **B**? *(4 marks)*

2 Suggest reagents, conditions and equations for each step in the synthetic routes below.

a $CH_3Br \rightarrow CH_3CN \rightarrow CH_3CH_2NH_2$ *(4 marks)*

b $C_6H_6 \rightarrow C_6H_5CHO \rightarrow C_6H_5CH(OH)CN \rightarrow C_6H_5CH(OH)COOH$ *(6 marks)*

3 Benzoic acid, C_6H_5COOH can be prepared as outlined below.

Stage 1: Methyl benzoate is heated for 30 minutes with NaOH(aq).

Stage 2: The reaction mixture is cooled and acidified with HCl(aq). Impure solid benzoic acid forms.

Stage 3: The impure benzoic acid is purified.

a Write equations for the reactions taking place in Stage 1 and Stage 2. *(2 marks)*

b **i** What procedure would you use for Stage 1? *(1 mark)*

ii List the Quickfit apparatus needed. *(2 marks)*

c At Stage 2, how could you test to know when the solution has been acidified? *(1 mark)*

d Outline how would you purify the benzoic acid. *(3 marks)*

e How would you expect the melting points of impure and pure benzoic acid to differ? *(2 marks)*

f In this synthesis, 5.28 g of methyl benzoate formed 3.76 g of benzoic acid. Calculate the percentage yield. *(3 marks)*

4 A student plans to synthesise two compounds from different starting materials.

a The first synthesis needs two stages:

$H_2NCH_2COOH \rightarrow$ intermediate $\rightarrow CH_3CONHCH_2COOCH_3$

i What are the functional groups in the starting and target molecules? *(2 marks)*

ii Devise a two-stage synthesis for this conversion.

Your answer should include reagents and conditions, and equations for each stage. *(4 marks)*

b The second synthesis needs three stages:

$CH_3CHO \rightarrow$ intermediate $\rightarrow$ intermediate $\rightarrow H_2C{=}CHCOOH$

i What are the functional groups in the starting and target molecules? *(2 marks)*

ii Devise a three-stage synthesis for this conversion.

Your answer should include reagents and conditions, and provide equations for each stage. *(6 marks)*

29.1 Chromatography and functional group analysis

Specification reference: 6.3.1

Chromatography

Chromatography is used to separate the components in a mixture.

Chromatography has a stationary phase and a mobile phase.

- The **stationary phase** does **not** move, and is normally a solid or a liquid on a solid support.
- The **mobile phase** does move, and is normally a liquid or a gas.

Thin layer chromatography (TLC)

- The stationary phase is a thin solid layer on the TLC plate. A sample binds to the surface of the stationary phase on the TLC plate by 'adsorption'.
- The mobile phase is a solvent, which moves up the TLC plate.

To run a TLC chromatogram:

- A solution of the sample is spotted onto the TLC plate using a capillary tube at the sample line.
- The TLC plate is placed in a solvent.
- The stronger the adsorption of a component to the solid stationary phase, the slower it moves up the TLC plate.

Analysing TLC chromatograms

If separation has been achieved, different components show up as different spots.

The value for the retention factor R_f of each component is calculated:

$$R_f = \frac{\text{distance moved by the component}}{\text{distance moved by the solvent front}}$$

Identifying an unknown compound

Figure 1 shows a developed TLC chromatogram for an unknown amino acid.

- The R_f value of the sample $= \frac{2.82}{4.63} = 0.61$
- The R_f value matches the amino acid leucine in Table 1.

Gas chromatography (GC)

Gas chromatography is used for separating and identifying volatile organic compounds in a mixture.

- The stationary phase is a high boiling point liquid adsorbed onto an inert solid support within a capillary column.
- The mobile phase is an inert carrier gas, e.g. helium.
- A small amount of the volatile mixture is injected into the gas chromatograph.
- The mobile carrier gas carries the components in the sample through the capillary column containing the stationary phase.
- The more soluble the component is in the liquid stationary phase, the slower it moves through the capillary column.

Analysing GC chromatograms

The compounds in the mixture reach the detector at different times ('retention times') depending on their interactions with the stationary phase.

- Retention time is the time taken for each component to travel through the column.

Revision tip

In TLC, the components of the mixture are separated by their relative adsorptions to the solid stationary phase.

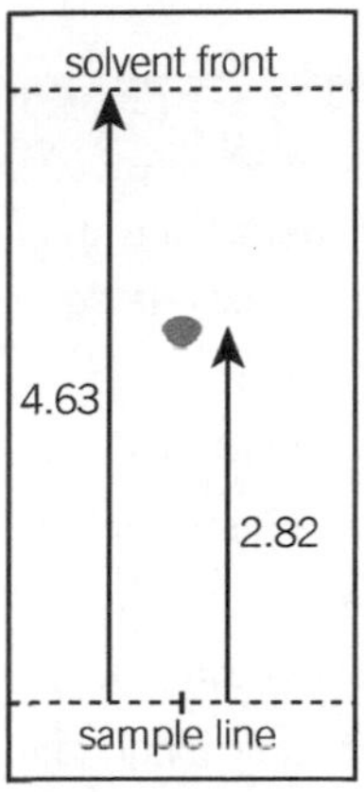

▲ **Figure 1** *TLC chromatogram of an amino acid*

▼ **Table 1** R_f values of amino acids

Amino acid	R_f value
aspartic acid	0.24
alanine	0.33
cysteine	0.37
valine	0.44
isoleucine	0.53
leucine	0.61

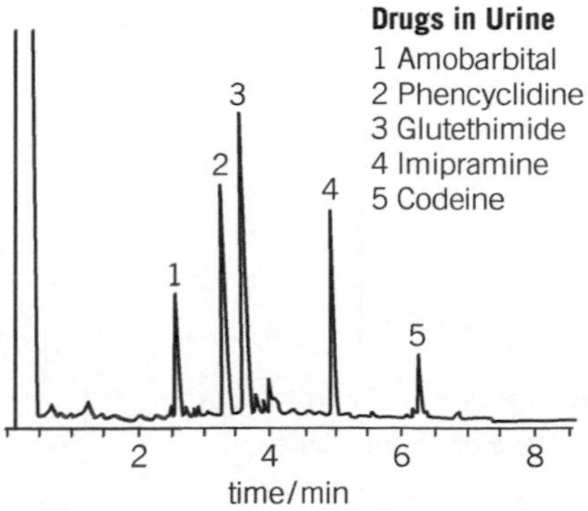

▲ **Figure 2** *Gas chromatogram of drugs in a urine sample*

Revision tip

In GC, the components of the mixture are separated by their relative solubility in the liquid stationary phase.

Figure 2 shows a gas chromatogram of a urine sample.

- The components have been identified from known retention times.
- You can get some idea of the amounts of the components by the relative size of the peaks.

Concentrations of components

The concentration of a component in a sample is determined by comparing its peak integration (peak area) with values obtained from standard solutions of the component.

Qualitative analysis of organic functional groups

You have already seen tests for different functional groups throughout the course. These tests are summarised in Table 2 below.

▼ **Table 2** *Tests for functional groups*

Functional group	Chemical test	Observation
Alkene	Add bromine water	bromine water decolourised from orange to colourless
Haloalkane	Add silver nitrate and ethanol Warm to 50 °C in a water bath	chloroalkane → white precipitate bromoalkane → cream precipitate iodoalkane → yellow precipitate
Carbonyl	Add 2,4-dinitrophenylhydrazine	yellow/orange precipitate
Aldehyde	Add Tollens' reagent and warm	silver mirror
Primary and secondary alcohol, and aldehyde	Add acidified potassium dichromate(VI) and warm	colour change from orange to green
Carboxylic acid	Add aqueous sodium carbonate	effervescence
Phenols	pH indicator paper and then add aqueous sodium carbonate	pH paper turns acid colour and no effervescence with $Na_2CO_3(aq)$

Synoptic link

For further detail, you should look at the Topics related to each functional group:

13.3, Reactions of alkenes,

15.1, The chemistry of the haloalkanes,

26.1, Carbonyl compounds,

26.2, Identifying aldehydes and ketones,

14.2, Reactions of alcohols,

26.3, Carboxylic acids, and

25.3, The chemistry of phenol.

Summary questions

1 a What is the mobile phase in TLC and in GC? *(2 marks)*

b How are components in a mixture separated by

i TLC *(1 mark)* **ii** GC? *(1 mark)*

2 a A mixture of three amino acids (**A**, **B**, **C**) is analysed by TLC (see Figure 3 and Table 1 from earlier in this topic).

Calculate the R_f values for the amino acids in the mixture and identify the amino acids. *(4 marks)*

b i Describe a chemical test that would distinguish between a carboxylic acid, RCOOH, and phenol. *(1 mark)*

ii Write an equation for the reaction taking place in **i**. *(1 mark)*

3 Compounds **A**, **B**, and **C** all turn warm acidified dichromate a green colour. The organic products of this test are added to aqueous sodium carbonate. The products from **A** and **B** effervesce but there is no effervescence from the product of **C**. Compound **A** forms an orange precipitate with 2,4-DNP.
What are the functional groups in **A**, **B**, and **C**?
Explain your reasoning. *(5 marks)*

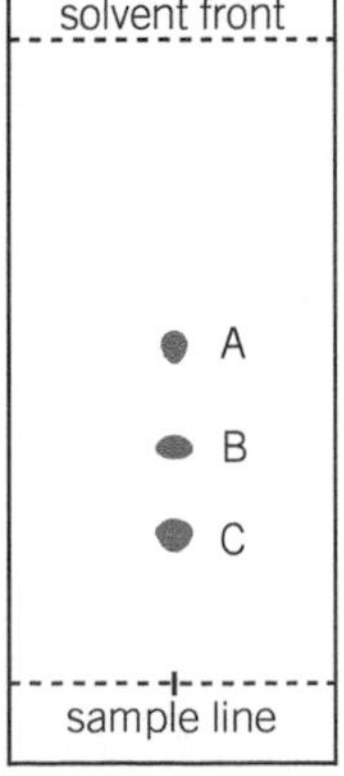

▲ **Figure 3** *TLC chromatogram for mixture of amino acids*

29.2 Nuclear magnetic resonance (NMR) spectroscopy

Specification reference: 6.3.2

Nuclear magnetic resonance (NMR)

NMR spectroscopy uses a combination of a very strong magnetic field and radio frequency radiation. The nuclei of some atoms absorb and release the radiation repeatedly in a process called nuclear magnetic resonance.

Carbon-13 and proton NMR spectroscopy

The nucleus of an atom contains nucleons (protons and neutrons).

- Only atoms with an odd number of nucleons absorb energy from radio waves.
- The ^{1}H isotope of hydrogen and ^{13}C isotope of carbon have an odd number of nucleons, giving carbon-13 (^{13}C) and proton (^{1}H) NMR spectroscopy.

Chemical shift

In an organic molecule, the electrons surrounding atoms cause a shift in the radio wave frequency that is absorbed when nuclear magnetic resonance takes place.

- This shift in frequency is measured on a scale called chemical shift, δ.
- The units of chemical shift, δ, are parts per million (ppm).

Tetramethylsilane (TMS), $(CH_3)_4Si$, is used as the standard for chemical shift measurements. The chemical shift of the ^{1}H and ^{13}C atoms in TMS is set at 0 ppm.

Chemical shift depends on the chemical environment of atoms and is greatly influenced by the presence of nearby electronegative atoms or π-bonds (see Figure 1).

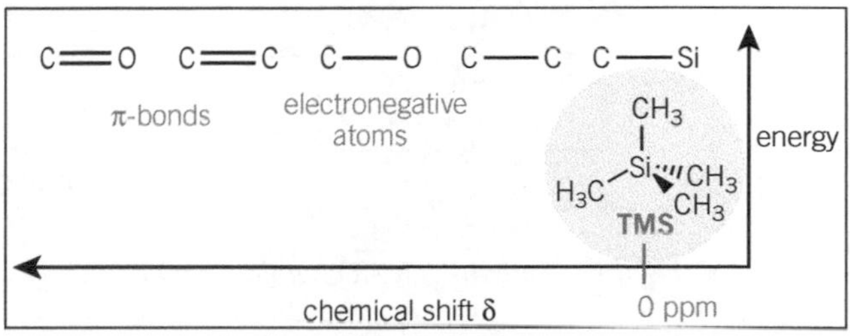

▲ **Figure 1** *The chemical shift scale*

Deuterated solvents

In NMR spectroscopy, the sample is tested in solution. Most organic solvents contain C and H atoms which produce peaks in both ^{13}C and ^{1}H NMR spectra.

Deuterium, D, is the ^{2}H isotope of hydrogen and has an even number of nucleons. Deuterated solvents such as $CDCl_3$ are used in NMR spectroscopy because they do **not** produce a ^{1}H peak in the spectrum.

Revision tip

Almost all organic molecules contain carbon (mostly as the ^{12}C isotope) and hydrogen (mostly as the ^{1}H isotope).

Carbon-13 makes up only 1.1% of carbon but modern NMR spectrometers are sensitive enough to detect absorption by ^{13}C.

Synoptic link

For details of isotopes and deuterium, see Topic 2.1, Atomic structure and isotopes.

Revision tip

In an NMR spectrum, the chemical shift in ppm increases from right to left.

The TMS peak, at 0 ppm, is at the right-hand end of the chemical shift scale.

Revision tip

A deuterated solvent has all H atoms present as the ^{2}H isotope.

Summary questions

1 Which of these isotopes produce peaks in an NMR spectrum?
^{1}H, ^{2}H, ^{12}C, ^{13}C, ^{14}N, ^{15}N, ^{16}O, ^{31}P, ^{32}P *(1 mark)*

2 a Give an example of a deuterated solvent and state why deuterated solvents are used. *(2 marks)*

b Dimethyl sulfoxide (DMSO), $(CH_3)_2SO$ is a good solvent for many organic compounds.

i Why is DMSO unsuitable for use as a solvent in ^{1}H NMR spectroscopy? *(1 mark)*

ii How could DMSO be modified to become a suitable NMR solvent? *(1 mark)*

3 a What is meant by chemical shift and what does it depend on? *(2 marks)*

b i What is the formula of TMS? *(1 mark)*

ii Why do all the protons in TMS absorb at the same chemical shift? *(1 mark)*

29.3 Carbon-13 NMR spectroscopy

Specification reference: 6.3.2

Revision tip

In a ^{13}C NMR spectrum, some peaks may be larger than others.

- The size of a peak is **not** directly related to the number of C atoms in that environment.

This is different from a ^{1}H NMR spectrum, studied in the next topics.

Revision tip

Chemical shifts are supplied for you on the data sheet.

Carbon-13 NMR spectroscopy

Analysis of a carbon-13 NMR spectrum provides two important pieces of information about a molecule:

- the *number* of different carbon environments – from the number of peaks
- the *types* of carbon environment present – from the chemical shift.

Chemical shifts for ^{13}C NMR spectroscopy

In ^{13}C NMR spectroscopy, chemical shift values of different carbon environments span a wide range of about 220 ppm (Figure 1).

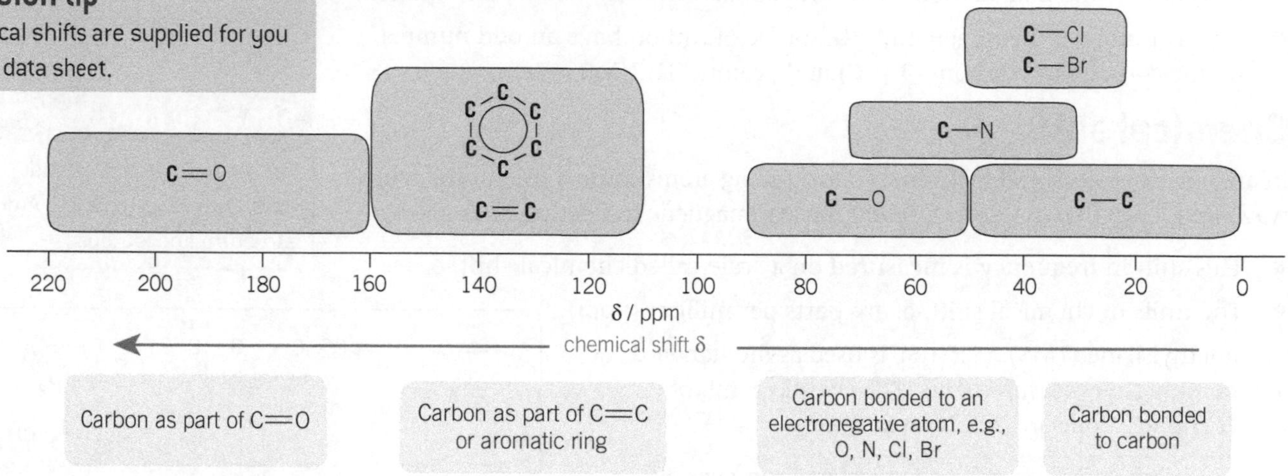

▲ **Figure 1** *Carbon-13 NMR chemical shifts*

Revision tip

It is important that you can recognise what makes different environments. Look for any plane of symmetry. This is a good way of visualising equivalent and non-equivalent carbon atoms.

Different environments and chemicals shifts

Carbon atoms with the **same** environment are **equivalent**.

Carbon atoms with **different** environments are **non-equivalent**.

Carbon atoms that are positioned symmetrically within a molecule

- are equivalent and have the same chemical environment
- have the same chemical shift and contribute to the same peak.

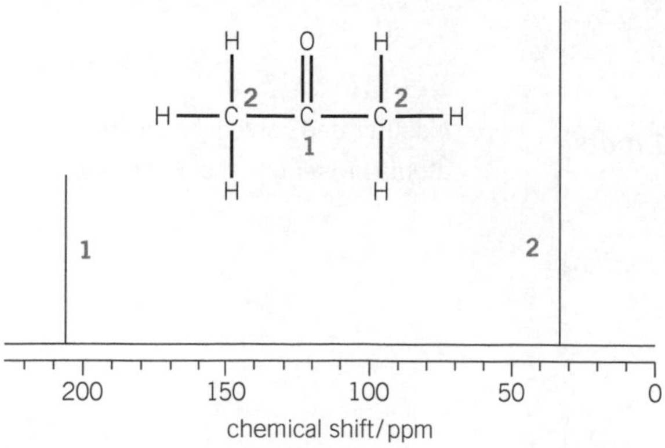

▲ **Figure 2** *Carbon-13 NMR spectrum of propanone*

The ^{13}C NMR spectrum of propanone, CH_3COCH_3

The ^{13}C NMR spectrum of CH_3COCH_3 (Figure 2) has two peaks, labelled 1 and 2, for two different carbon environments.

There is a vertical plane of symmetry through the C=O bond of the molecule.

- The C=O group gives one peak.
- The two CH_3 groups are positioned symmetrically, have the same environment, and contribute to the same peak.

Chemical shifts

You can identify the type of carbon environment by matching the chemical shift of a peak with known chemical shifts (see Figure 1).

In the NMR spectrum in Figure 2:

- Peak-1 is at δ ~ 205 ppm for the C atom of type C=O.
- Peak-2 is at δ ~ 32 ppm for the two equivalent C atoms (of type C–C) in the two CH_3 groups.

Predictions for possible structures of a molecule

You can predict a possible structure for a molecule from its ^{13}C NMR spectrum.

The Worked example shows how ^{13}C NMR spectra can be used to predict the substitution positions in a multisubstituted aromatic compound.

Worked example: Identifying a structure from a ^{13}C NMR spectrum

How can the aromatic isomers of C_8H_{10} be identified from their ^{13}C NMR spectra?

Step 1: Draw out the structures of the possible aromatic isomers of C_8H_{10}.

Identify the number of carbon environments in each structure (see the numbers).

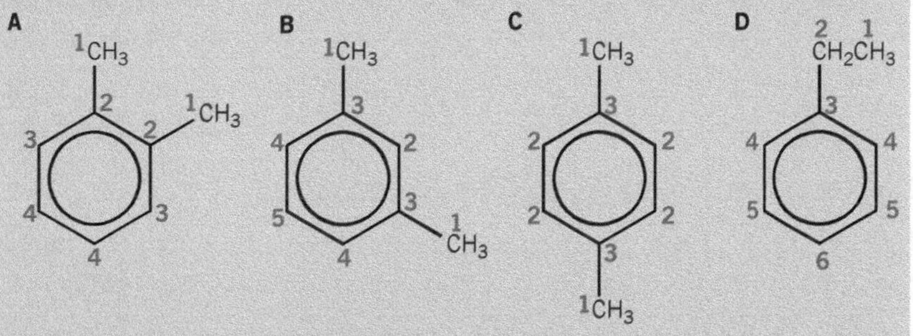

▲ **Figure 3** *Number of carbon environments in the aromatic isomers of C_8H_{10}*

Step 2: Identify the isomers from the number of peaks in the ^{13}C NMR spectrum.

The number of peaks for each structure is equal to the number of carbon environments (see Figure 3).

The isomers can be identified from the number of peaks:

Number of peaks	4	5	3	6
Isomer	A	B	C	D

Revision tip

Structure A has a plane of symmetry going midway between the two C atoms labelled '2' and the two C atoms labelled '4'.

Check each structure so that you can see how the number of carbon environments has been decided. The number of peaks would be different for all four isomers.

Summary questions

Use the ^{13}C chemical shifts in Figure 1 to help you answer these questions.

1 For each compound, predict the number of ^{13}C peaks and the chemical shift of each peak:

a $CH_3CH(OH)CH_3$ *(2 marks)*
b CH_3COOCH_3 *(2 marks)*
c $CH_3CH_2COCH_2CH_3$ *(2 marks)*
d C_6H_5OH *(2 marks)*

2 **A** and **B** are structural isomers of C_4H_9Cl. In their ^{13}C NMR spectra, **A** has 3 peaks and **B** has 2 peaks.

Predict the number of peaks in the four structural isomers of C_4H_9Cl and identify **A** and **B**. *(5 marks)*

3 Trichlorophenol has 2 isomers with 4 peaks in their ^{13}C NMR spectra.

Draw the structure of each isomer and label each carbon environment. *(4 marks)*

29.4 Proton NMR spectroscopy

Specification reference: 6.3.2

Proton NMR spectroscopy

Analysis of a proton NMR spectrum provides **four** important pieces of information about a molecule:

- the *number* of different proton environments – from the number of peaks
- the *types* of proton environment present – from the chemical shift
- the *relative numbers* of each type of proton – from relative peak areas or integration traces
- the number of *adjacent* non-equivalent protons – from the spin–spin splitting pattern.

Revision tip

Chemical shifts are supplied for you on the data sheet.

Chemical shifts for ^1H NMR spectroscopy

In ^1H NMR spectroscopy, chemical shift values span a range of about 12 ppm, much narrower than for ^{13}C. You may find that some peaks overlap.

Figure 1 shows chemical shifts for protons in different environments.

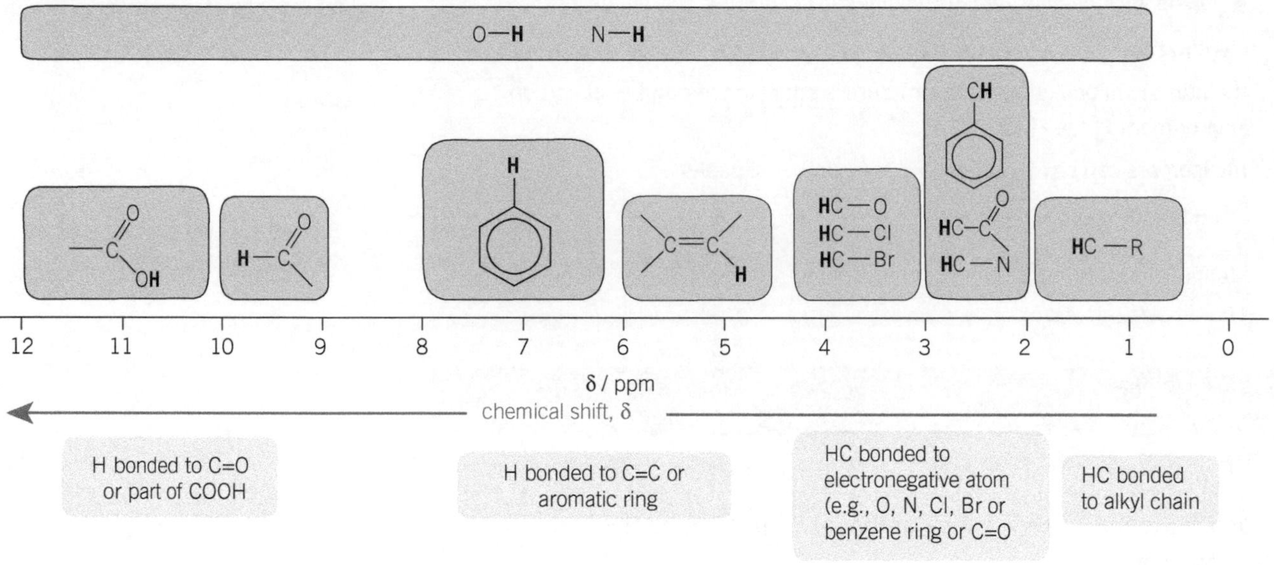

▲ **Figure 1** *Proton NMR chemical shifts*

Number of protons in each environment

In a ^{13}C NMR spectrum, the peak area is **not** directly related to the number of C atoms responsible for the peak.

This is different for ^1H NMR:

- the peak area increases by a set amount for each additional proton.
- the ratio of the peak areas gives the ratio of the number of protons responsible for each peak.

A ^1H NMR spectrum can show peak areas as an 'integration trace'. Figure 2 shows the integration trace in the ^1H NMR spectrum of methyl chloroethanoate, $ClCH_2COOCH_3$. You can see the peak area ratio of 2 : 3 for the two protons in CH_2 and three protons in CH_3.

Often, the relative peak areas are simply shown as a number on a ^1H NMR spectrum.

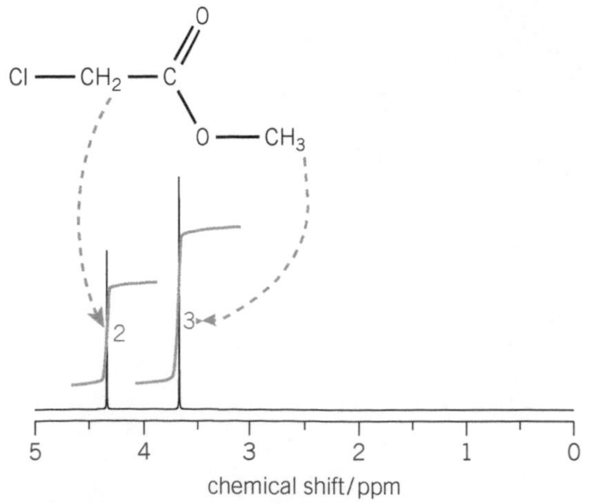

▲ **Figure 2** *Integration trace for methyl chloroethanoate, $ClCH_2COOCH_3$*

Different environments, chemicals shifts, and peak areas

The chemical environments and expected peak areas for two organic molecules are described below.

Butanoic acid, $CH_3CH_2CH_2COOH$ (Figure 3)

There is no plane of symmetry and the 8 protons can be divided into four different environments, labelled in Figure 3 as 1–4.

- 3 protons (1) in the CH_3 group.
- 2 protons (2) in the CH_2 group positioned between a CH_3 group and CH_2.
- 2 protons (3) in the CH_2 group positioned between a CH_2 group and CO.
- 1 proton (4) in the COOH group.

The four proton environments produce a ^1H NMR spectrum with four peaks.

The relative peak areas would be 3 : 2 : 2 : 1 for CH_3, CH_2, CH_2, COOH.

Butanedioic acid, $HOOCCH_2CH_2COOH$ (Figure 4)

There is a plane of symmetry and the 6 protons can be divided into two different environments, labelled in Figure 4 as 1–2.

- 4 protons (1) in two equivalent CH_2 groups.
- 2 protons (2) in two equivalent COOH groups.

The two proton environments produce an NMR spectrum with two peaks.

The relative peak areas would be 2 : 1 for 4H ($2 \times CH_2$) : 2H ($2 \times$ COOH).

Revision tip

For both ^{13}C NMR and ^1H NMR, look for any plane of symmetry. This is a good way of visualising equivalent and non-equivalent protons.

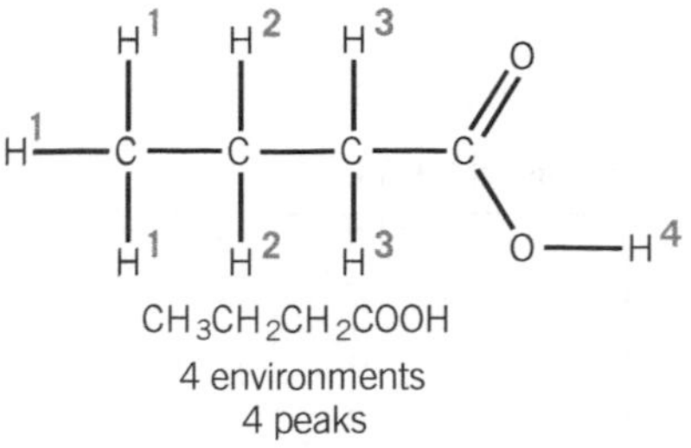

▲ **Figure 3** *$CH_3CH_2CH_2COOH$ has 4 carbon environments*

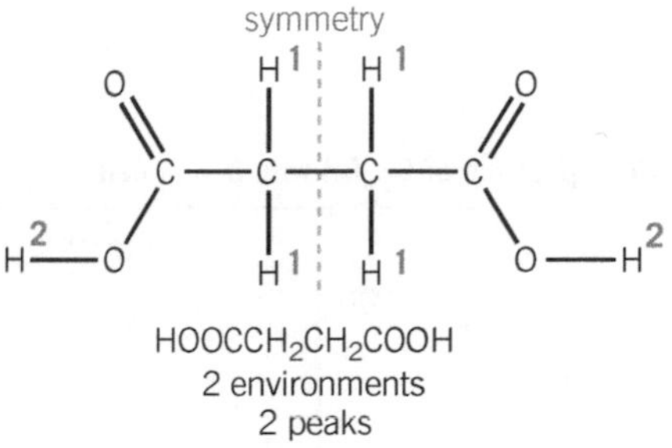

▲ **Figure 4** *$HOOCCH_2CH_2COOH$ has 2 proton environments*

Spin–spin splitting

A ^1H NMR peak may be split into sub-peaks by 'spin–spin splitting'. The splitting is caused by interactions between adjacent non-equivalent protons (in different environments).

The $n + 1$ rule

The splitting pattern is easily worked out using the $n + 1$ rule:

- For n protons on an adjacent carbon atom, splitting pattern = $n + 1$.

Table 1 summarises the common splitting patterns.

▼ **Table 1** *Spin–spin splitting patterns*

n	$n+1$	Splitting pattern	Relative peak areas within splitting	Pattern	Structural feature
0	1	singlet	1		no H on adjacent atoms
1	2	doublet	1 : 1		adjacent CH
2	3	triplet	1 : 2 : 1	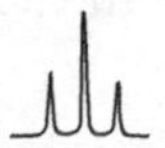	adjacent CH_2
3	4	quartet	1 : 3 : 3 : 1	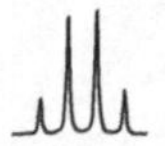	adjacent CH_3

You may also see a heptet, seven sub-peaks, caused by the six protons in two adjacent CH_3 groups, as in $CH(CH_3)_2$ (Figure 5).

Except for a singlet, splitting always comes in pairs.

If you see one splitting pattern there must be another.

Revision tip

Spin–spin splitting only occurs if adjacent protons are in a different environment from the protons being split. If you are unsure, look back to the examples of equivalent and non-equivalent protons earlier in this topic.

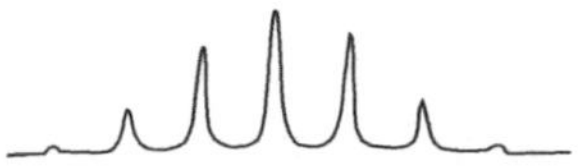

▲ **Figure 5** *Heptet splitting of the CH group in $CH(CH_3)_2$, caused by the six adjacent protons in the two CH_3 groups*

Some splitting patterns are very common, e.g.

- A triplet and a quartet show a CH_3CH_2 group. Relative peak areas 3 : 2
- A heptet and doublet show a $CH(CH_3)_2$ group. Relative peak areas 1 : 6

Aromatic protons

From Figure 1, aromatic protons are expected to absorb in the range δ = 6.2–8.0 ppm (see Figure 1). There are splitting patterns but these go beyond A level Chemistry.

Identification of O–H and N–H protons

There are problems in assigning O–H and N–H protons in an NMR spectrum.

The NMR peaks are often:

- broad (from hydrogen bonding) and have no splitting pattern
- of variable chemical shift (see Figure 1).

Carboxylic acid COOH protons are more predictable, absorbing at 10–12 ppm.

Proton exchange using deuterium oxide, D_2O

Proton exchange with D_2O is used for identifying O–H and N–H protons:

1. A ^1H NMR spectrum is run as normal.
2. A few drops of D_2O are added, the mixture is shaken, and a second spectrum is run.

Deuterium exchanges with any OH and NH protons.

When the second spectrum is run, O–H and N–H peaks disappear, allowing the peaks to be assigned.

Figure 6 shows the NMR spectra of methanol without and with D_2O added.

NMR spectrum of CH_3OH

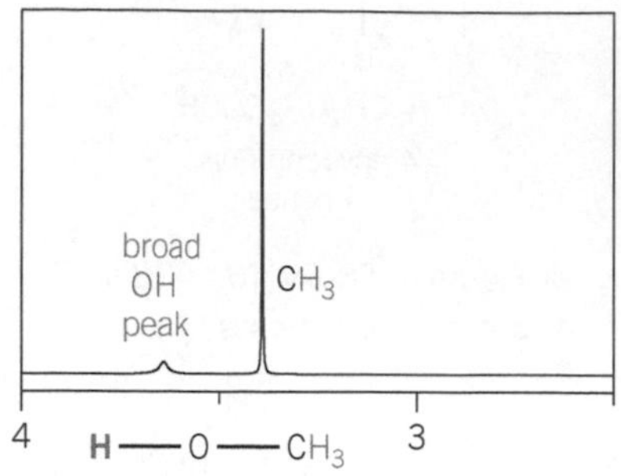

NMR spectrum of CH_3OH with D_2O added

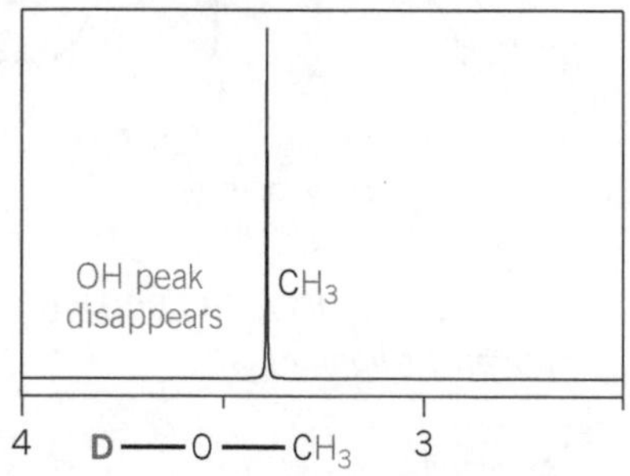

▲ **Figure 6** *NMR spectra of methanol, CH_3OH, run without and with D_2O*

Synoptic link

In Topic 29.2, Nuclear magnetic resonance (NMR) spectroscopy, you came across deuterated solvents.

Summary questions

Use the proton chemical shifts given in Figure 1 to help you answer these questions.

1. For each compound, predict the number of peaks, the relative peak areas, and any difference after D_2O has been added.
   - a $CH_3CH_2CH_2CHO$ *(3 marks)*
   - b $(CH_3)_2CHCH_2OH$ *(3 marks)*
   - c CH_3NHCH_3 *(3 marks)*
   - d H_2NCH_2COOH *(3 marks)*
2. For each compound, predict the spin–spin splitting of each peak.
   - a CH_3CH_2OH *(3 marks)*
   - b $HOCH_2CH_2OH$ *(2 marks)*
   - c $(CH_3)_2CHNH_2$ *(3 marks)*
   - d $HOCH_2CH_2COOCH_3$ *(4 marks)*
3. Write the groups that produce the following splitting patterns in a ^1H NMR spectrum:
   - a triplet and quartet *(1 mark)*
   - b triplet and doublet *(1 mark)*
   - c doublet and doublet *(1 mark)*
   - d doublet and heptet *(1 mark)*

29.5 Interpreting NMR spectra

Specification reference: 6.3.2

Analysis of ^{1}H NMR spectra of an organic molecule

There are no set rules for identifying a compound from a proton NMR spectrum. The Worked examples follow a step-by-step procedure but you can solve spectral problems in almost any order. It depends on which piece of evidence you see first and there are countless ways of solving these problems.

Worked example: Analysing a ^{1}H NMR spectrum

An isomer of $C_3H_6O_3$ gives the ^{1}H NMR spectrum in Figure 1. The numbers are the relative peak areas. The peaks at $\delta = 11.0$ ppm and $\delta = 2.8$ ppm disappear after addition of D_2O.

Analyse the spectrum to identify the compound.

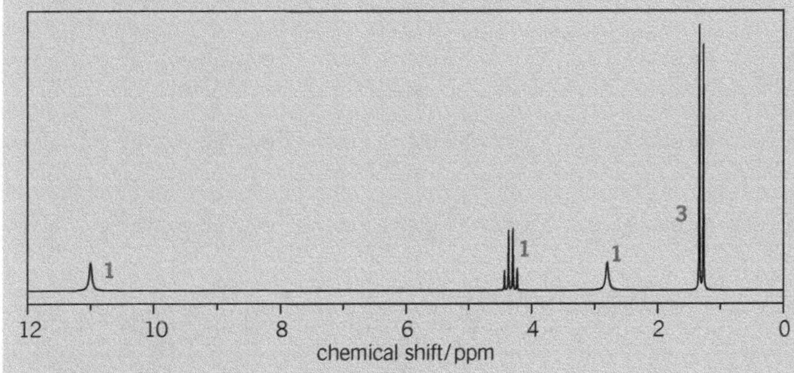

▲ **Figure 1** *^{1}H NMR spectrum of isomer of $C_3H_6O_3$*

Step 1: Analyse the types of proton present and the number of each type.

- Peaks at 11.0 ppm and 2.8 ppm disappear with D_2O → OH peaks.
- Four peaks → four types of proton environment.
- Relative peak areas → H ratio = 1 : 1 : 1 : 3 (from left) → OH : CH : OH : CH_3.

Step 2: Analyse the splitting patterns to find information about adjacent protons.

- CH quartet at 4.2 ppm $n + 1 = (3 + 1) = 4$ adjacent CH_3
- CH_3 doublet at 1.2 ppm $n + 1 = (1 + 1) = 2$ adjacent CH

} CH_3CH

Step 3: Use the data in Topic 29.4, Figure 1 to analyse the chemical shifts.

- OH peak at $\delta = 11.0$ ppm type COOH → –COOH
- CH peak at $\delta = 4.2$ ppm type HC–O → –CH–OH
- OH peak at $\delta = 2.8$ ppm type OH → –OH
- CH_3 peak at $\delta = 1.2$ ppm type HC–R → CH_3–C

Step 4: Combine the information to suggest a structure.

- The correct structure must be $CH_3CH(OH)COOH$. Figure 2 shows the four proton environments.

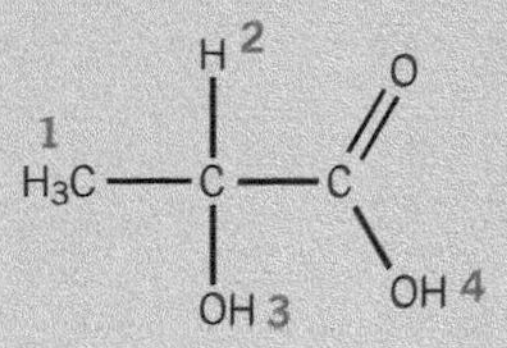

▲ **Figure 2** *Four proton environments in $CH_3CH(OH)COOH$*

Synoptic link

For details of proton NMR spectra, see Topic 29.4, Proton NMR spectroscopy.

Revision tip

A peak that disappears with D_2O must be for OH, COOH or NH.

Revision tip

Chemical shift values for different types of proton environments are provided on the data sheet.

Synoptic link

For details of predicting a carbon-13 NMR spectrum, see Topic 29.3, Carbon-13 NMR spectroscopy.

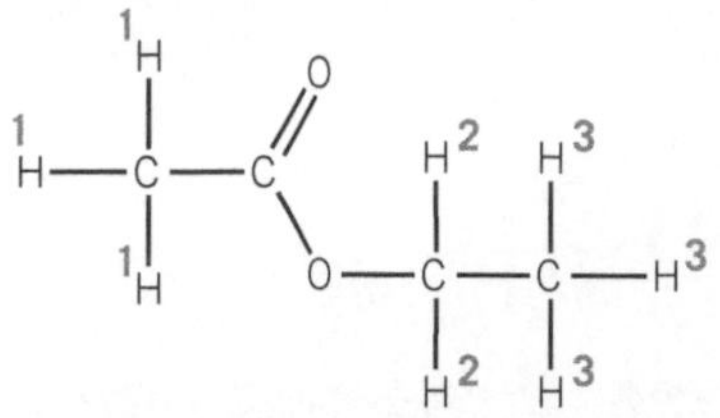

▲ **Figure 3** *Proton environments in $CH_3COOCH_2CH_3$*

Revision tip

$CH_3COOCH_2CH_3$ has the common CH_3CH_2 sequence which gives the triplet–quartet splitting pattern.

Predicting NMR spectra

Using the information from Topics 29.3 and 29.4, you can predict both ^{13}C and ^{1}H NMR spectra.

Worked example: Predicting a ^1H NMR spectrum

Predict the ^1H NMR spectrum for $CH_3COOCH_2CH_3$.

Step 1: Draw out the structure and identify the number of proton environments.

- There are **three** proton environments (H1–H3) giving **three** peaks.

Step 2: Predict the relative peak areas from the number of each type of proton.

- Relative peak areas (H1–H3) → 3 : 2 : 3 (CH_3, CH_2, CH_3)

Step 3: Predict the splitting patterns from the number of adjacent protons.

- H1 CH_3 0 Hs on adjacent C=O $n + 1 = (0 + 1) = 1$ → singlet
- H2 CH_2 **3 Hs on adjacent CH_3** $n + 1 = (\mathbf{3} + 1) = 4$ → quartet
- H3 CH_3 **2 Hs on adjacent CH_2** $n + 1 = (\mathbf{2} + 1) = 3$ → triplet

Step 4: Use the data in Topic 29.4, Figure 1, to predict the chemical shifts.

- H1 3 protons, CH_3 type **HC–CO** $\delta = 2.0–3.0$ ppm (singlet)
- H2 2 protons, CH_2 type **HC–O** $\delta = 3.0–4.2$ ppm (quartet)
- H3 3 protons, CH_3 type **HC–R** $\delta = 0.5–2.0$ ppm (triplet)

Summary questions

1 An alcohol, $C_5H_{10}O$, has the three peaks in its ^{13}C NMR spectrum, at 25 ppm, 34 ppm, and 78 ppm. Include your reasoning. Identify the compound. *(3 marks)*

2 Identify the following compounds from their formula and ^1H NMR peaks. Include your reasoning.

a An isomer of $C_4H_8O_2$ with three peaks: a triplet (3H) at 1.2 ppm, a quartet (2H) at 2.2 ppm, and a singlet (3H) at 3.8 ppm. *(4 marks)*

b An isomer of $C_3H_6O_3$ with four peaks: a triplet (2H) at 2.2 ppm, a triplet (2H) at 3.8 ppm and two 1H singlets at 4.5 ppm and 11.5 ppm, which both disappear with D_2O. *(5 marks)*

c An isomer of $C_6H_{12}O$ with three peaks: a singlet (9H) at 1.2 ppm, a doublet (2H) at 2.2 ppm and a triplet (1H) at 9.5 ppm. *(4 marks)*

3 An isomer of $C_5H_{10}O$ has the ^1H NMR spectrum in Figure 4. The numbers are the relative peak areas. The peak at 2.7 ppm is a heptet. Analyse the spectrum to identify the compound. Include your reasoning. *(4 marks)*

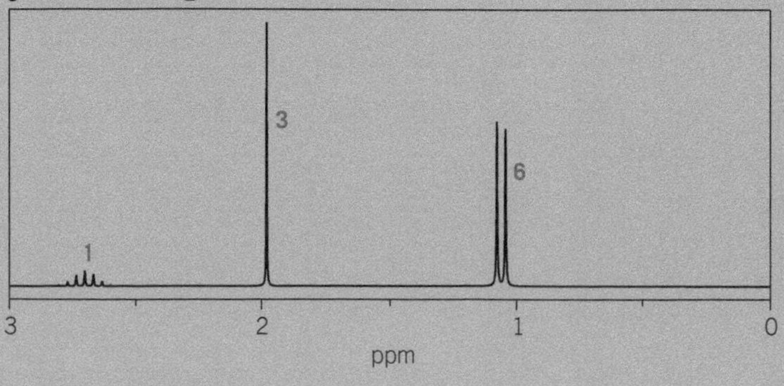

▲ **Figure 4** *^1H NMR spectrum of $C_5H_{10}O$*

29.6 Combined techniques

Specification reference: 6.3.2

Structure determination

Organic chemists use different analytical information to determine the structure of an organic molecule:

- Elemental analysis → empirical formula
- Mass spectra → molecular mass and fragments of structure
- IR spectra → bonds present and functional groups
- NMR spectra → environments and structural formula.

Synoptic link

If you need to revise these techniques, see:

- Topic 3.2, Determination of formulae, for empirical formula and molecular calculations.
- Topic 17.1, Mass spectrometry, for molecular ion peak, and fragment ions.
- Topic 17.2, Infrared spectroscopy, for IR spectra.
- Topics 29.2–29.5 for NMR spectroscopy.

Worked example: Analysing a compound by combined techniques

Analyse the evidence to suggest the structural formula of the unknown compound.

Elemental analysis by mass: C, 54.55%; H, 9.09%; O, 36.36%.

Mass, IR, and ^{13}C and ^{1}H NMR spectra are shown in Figures 1–4.

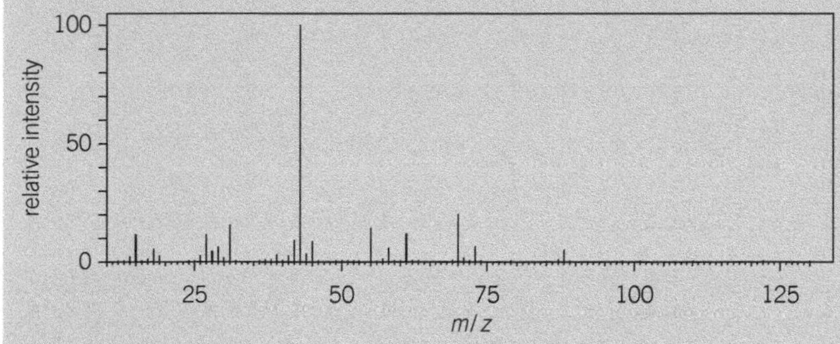

▲ **Figure 1** *Mass spectrum of unknown compound*

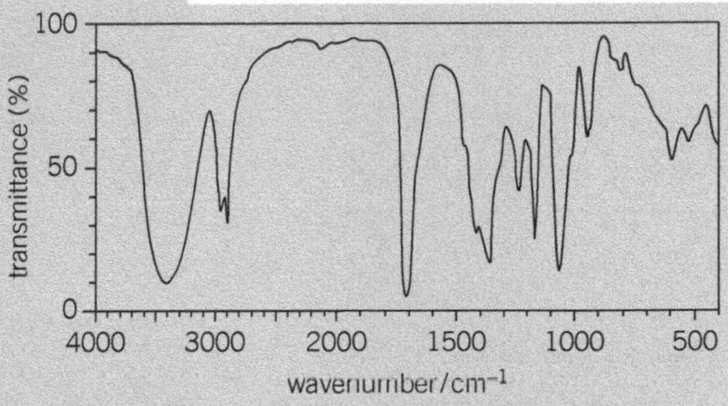

▲ **Figure 2** *IR spectrum of unknown compound*

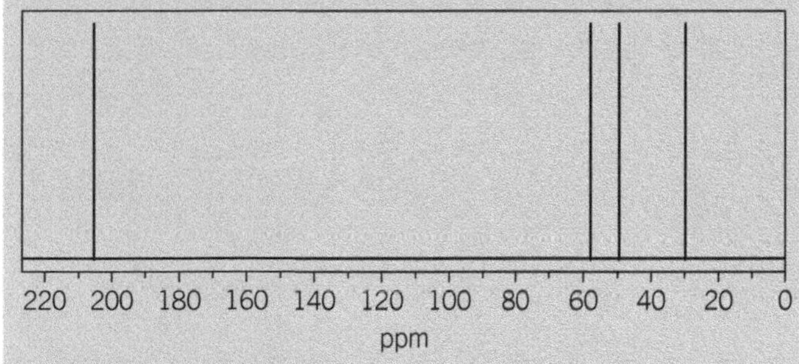

▲ **Figure 3** *^{13}C NMR spectrum of unknown compound*

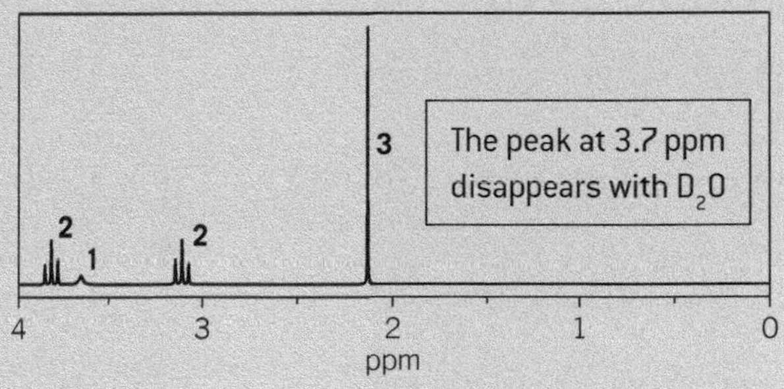

▲ **Figure 4** *Proton NMR spectrum of unknown compound*

Step 1: Determine the empirical formula from the elemental analysis data.

Convert % by mass to amounts in moles.

$$n(C) = \frac{54.55}{12.0} = 4.55\text{ mol} \quad n(H) = \frac{9.09}{1.0} = 9.09\text{ mol} \quad n(O) = \frac{36.36}{16.0} = 2.27\text{ mol}$$

Find smallest whole number ratio and the empirical formula

$$n(C) : n(H) : n(O) = \frac{4.55}{2.27} : \frac{9.09}{2.27} : \frac{2.27}{2.27} = 2 : 4 : 1 \quad \text{Empirical formula} = C_2H_4O$$

Step 2: Determine the molecular formula using the mass spectrum and the empirical formula.

Molecular ion peak at $m/z = 88$ → molecular mass = 88

Relative mass of empirical formula = $12.0 \times 2 + 1.0 \times 4 + 16.0 \times 1 = 44.0$

$$\text{Molecular formula} = C_2H_4O \times \frac{88}{44} = C_4H_8O_2$$

Synoptic link

See Topic 3.2, Determination of formulae, for details of empirical and molecular formula calculations.

Synoptic link

See Topic 17.1, Mass spectrometry, for details of mass spectrometry.

Revision tip

Take care. The OH absorption is too high for a carboxylic acid OH.

The peak at 1710 cm^{-1} cannot be a carboxylic acid C=O as there is no broad peak present at 2500–3300 cm^{-1} for the carboxylic acid O–H.

∴ the compound contains a C=O and an alcohol OH.

Revision tip

In a ^1H NMR spectrum, first look for:

- the number of proton environments from the number of peaks
- the number of each type of proton from the relative peak areas.

Also check for any OH or NH protons from D_2O information.

Revision tip

Once you know the number of each type of proton, look at the ^1H NMR spectrum for:

- the adjacent protons from splitting patterns
- the types of proton from chemical shift values.

Step 3: Identify the functional groups using the IR spectrum and data sheet.

Peak at 1710 cm^{-1} → C=O group in an aldehyde, ketone, or ester.

Peak at 3400 cm^{-1} → O–H group in an alcohol.

Step 4: Analyse the ^{13}C NMR spectrum.

2 peaks at 0–50 ppm for C–C environment.

1 peak at 50–90 ppm for C–O environment.

1 peak at 160–220 ppm for C=O environment.

Step 5: Analyse the ^1H NMR spectrum for the number of different types of proton.

Peak at 3.7 ppm disappears with D_2O → OH peak.

Four peaks → four types of proton.

Relative peak areas → 2 : 1 : 2 : 3 (from left) → CH_2 : OH : CH_2 : CH_3.

Step 6: Analyse the ^1H NMR spectrum for adjacent protons and types of proton.

CH_2 triplet at 3.8 ppm	CH_2 adjacent to $-CH_2$ and –O:	$O-CH_2-CH_2$
CH_2 triplet at 3.1 ppm	CH_2 adjacent to $-CH_2$ and –C=O:	$CH_2-CH_2-C=O$
CH_3 singlet at 2.2 ppm	CH_3 adjacent to C=O:	$CH_3-C=O$

Step 7: Combine the information to propose a structural formula.

- Structural formula: $\mathbf{HOCH_2CH_2COCH_3}$.

Summary questions

Elemental analysis of an unknown compound gives the percentage composition by mass: C, 58.83%; H, 9.80%; O, 31.37%. The molecular ion peak is at m/z = 102. The IR and ^1H NMR spectra are shown in Figures 5–6.

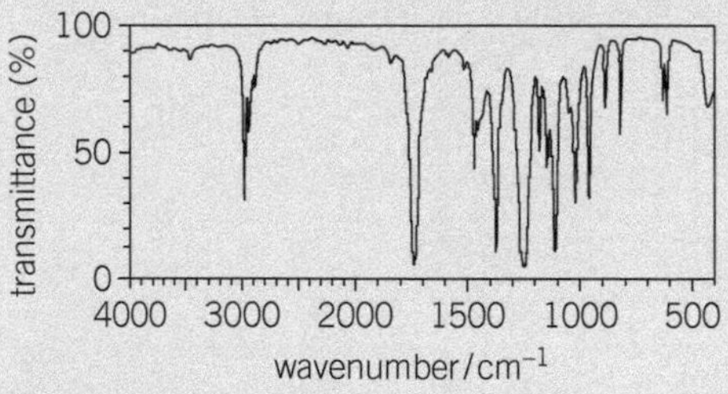

▲ **Figure 5** *The infrared spectrum of unknown compound*

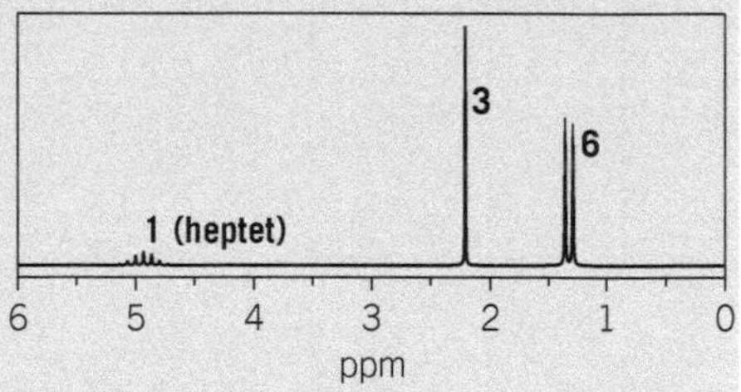

▲ **Figure 6** *^1H NMR spectrum of unknown compound*

1 Determine the empirical and molecular formulae of the compound. *(3 marks)*

2 Analyse the infrared spectrum. *(2 marks)*

3 **a** Analyse the ^1H NMR spectra to suggest a possible structure for the compound. *(6 marks)*

b Predict the ^{13}C NMR spectrum of the compound. *(4 marks)*

Chapter 29 Practice questions

1 An unknown compound forms an orange precipitate with 2,4-DNP but does **not** react with warm acidified dichromate(VI). The compound decolourises bromine water.

What is a possible structure for the unknown compound?

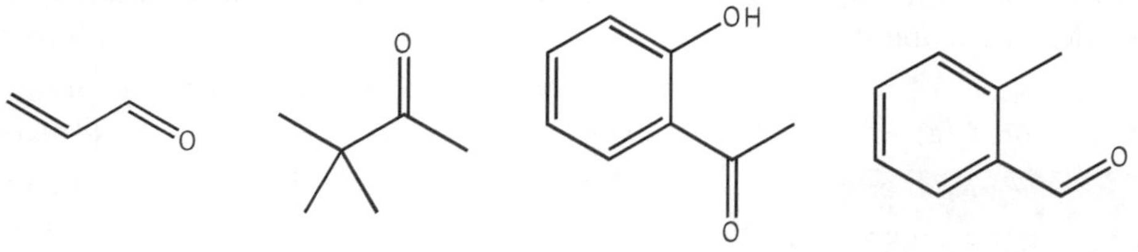

A **B** **C** **D**

(1 mark)

2 How many peaks are in a ^{13}C NMR spectrum of 1,3-dinitrobenzene?

A 3 **B** 4 **C** 5 **D** 6 *(1 mark)*

3 How many peaks are in a ^{1}H NMR spectrum of $HOCH_2C(CH_3)_2CH_2OH$?

A 3 **B** 4 **C** 5 **D** 6 *(1 mark)*

4 Which group produces a triplet and doublet in a ^{1}H NMR spectrum?

A CH_3CH_2 **B** CH_3CH

C CH_2CH **D** $(CH_3)_2CH$ *(1 mark)*

5 Compound **A** is an organic compound containing C, H, and O only.

Elemental analysis of compound **A** gives the following percentage composition by mass:

C, 64.62%; H, 10.77%; O, 24.61%.

The mass spectrum shows a molecular ion peak at $m/z = 130$.

The ^{1}H NMR and IR spectra are shown below.

Analyse the evidence to propose a structure for compound **A**. *(10 marks)*

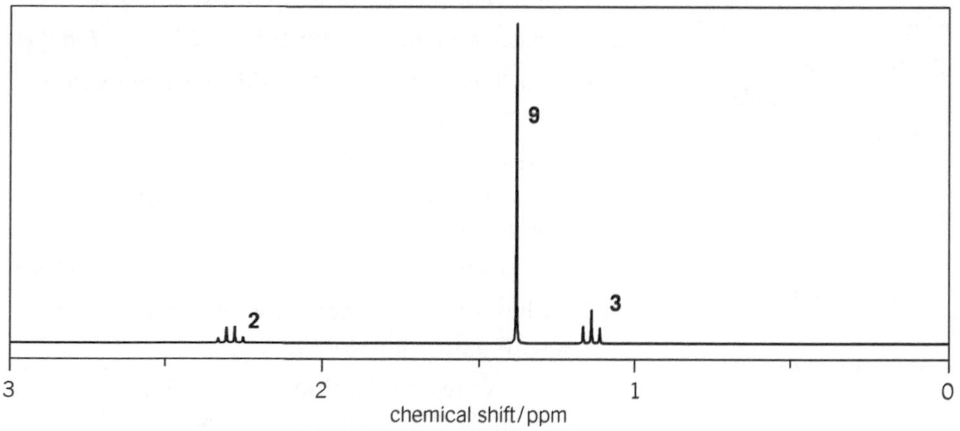

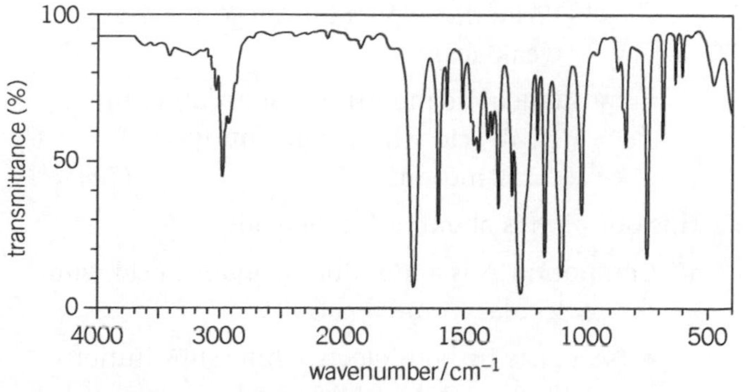

Synoptic questions

1 This question is about nitric acid, HNO_3.

a Nitric acid is produced by industry in three stages.

Stage 1: Ammonia is first heated with oxygen to form nitrogen monoxide, NO, at about 900 K:

$$4NH_3(g) + 5O_2(g) \rightleftharpoons 4NO(g) + 6H_2O(g)$$

	$NH_3(g)$	$O_2(g)$	$NO(g)$	$H_2O(g)$
$S^\ominus$ / J mol^{-1} K^{-1}	+192.3	+205.0	+210.7	+188.7
$\Delta_f H^\ominus$ / kJ mol^{-1}	−46.1	0	+90.2	−241.8

Stage 2: The nitrogen monoxide gas is cooled under pressure and reacted with air:

$$2NO(g) + O_2(g) \rightleftharpoons 2NO_2(g) \quad \Delta H = -114\ \text{kJ mol}^{-1}$$

Stage 3: The nitrogen dioxide is reacted with water to form a mixture of nitric acid, HNO_3, and nitrous acid, HNO_2.

i For **Stage 1**, calculate the free energy change, ΔG, in kJ mol^{-1}, at 900 K.

Show all your working. *(6 marks)*

ii For **Stage 2** explain why the mixture is cooled to 780 K under pressure. *(3 marks)*

iii At 780 K, an equilibrium mixture for **Stage 2** contains 1.80 mol NO, 2.10 mol O_2, and 8.10 mol NO_2. The total pressure is 10.5 atm.

Calculate K_p for this equilibrium at 780 K. *(6 marks)*

iv For **Stage 3**, write the equation for this reaction and show that disproportionation has taken place. *(3 marks)*

b Nitric acid can behave as an oxidising agent. Different redox reactions take place between magnesium and nitric acid, depending on the concentration of the nitric acid.

i Very dilute nitric acid reacts with magnesium to form hydrogen gas.

Write the equation for this reaction. *(1 mark)*

ii Concentrated nitric acid reacts with magnesium to form magnesium nitrate, nitrogen dioxide, and one other product.

Write the equation for this reaction. *(2 marks)*

2 This question is about compounds of sodium, chlorine, and oxygen.

a $NaClO_3$ contains Na^+ and ClO_3^- ions.

$NaClO_3$, is completely decomposed by strong heat to form **A** and **B**.

Compound **A** has the percentage composition by mass:

Na, 18.78%; Cl, 28.98; O, 52.24%.

Compound **B** forms a white precipitate with aqueous silver nitrate.

i What is the systematic name of $NaClO_3$? *(1 mark)*

ii Determine the formulae of **A** and **B**. Show your working and reasoning. *(4 marks)*

iii Write an equation for the decomposition of $NaClO_3$. *(1 mark)*

b Sulfur dioxide gas is bubbled into an aqueous solution of $NaClO_3$. A reaction takes place between ClO_3^- ions, SO_2 and H_2O to form an acid solution containing $Cl^-(aq)$ and $SO_4^{2-}(aq)$ ions.

Write the overall equation for this reaction. *(2 marks)*

c $NaClO_2$ contains Na^+ and ClO_2^- ions.

In the ClO_2^- ion, the Cl atom bonds to one O atom with a double covalent bond and to one O atom with a dative covalent bond. The Cl atom is surrounded by a total of 10 electrons.

i Draw a *'dot-and-cross'* diagram of a ClO_2^- ion. *(2 marks)*

ii Predict the shape and bond angles in a ClO_2^- ion.

Explain your reasoning. *(4 marks)*

d NaClO contains Na^+ and ClO^- ions.

i On heating, NaClO disproportionates, forming $NaClO_3$ and one other product.

Write an equation for this reaction. *(1 mark)*

ii NaClO can be used to oxidise cyclohexanol to cyclohexanone.

Write an equation for this reaction. Use structures for organic compounds and NaClO rather than [O] for the oxidising agent. *(2 marks)*

iii ClO^- is the conjugate base of a weak acid with a pK_a value of 7.53.

- Write an equation to show the dissociation of the weak acid.
- Write the expression for K_a for the weak acid.
- Calculate the pH of a solution of the weak acid with a concentration of 0.250 mol dm^{-3}. *(5 marks)*

3 This question is about different acids.

a Compound **A** is a monobasic organic acid with the molecular formula $C_xH_yO_2$.

- **A** reacts by both electrophilic substitution and electrophilic addition.

- The mass spectrum of **A** has a molecular ion peak at $m/z = 148$.
- **A** is an *E* stereoisomer.

Determine and draw a possible structure for **A**. Explain your reasoning. *(4 marks)*

b Compounds **C**, **D**, **E**, and **F**, shown in Table 1, are structural isomers.

▼ **Table 1** *Structural isomers, C, D, E, and F*

	Formula	Melting point/°C
C	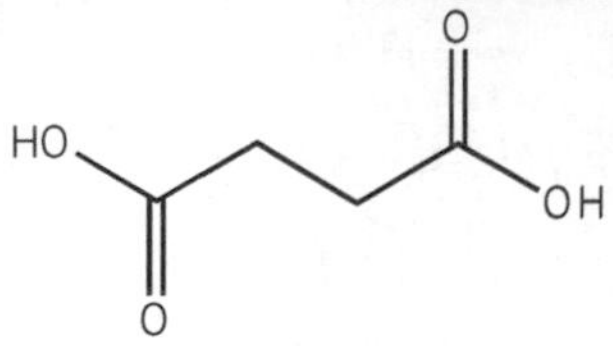	187
D	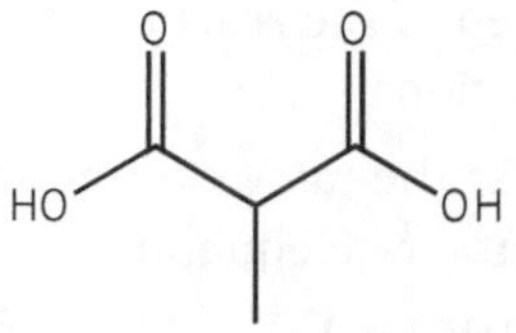	136
E	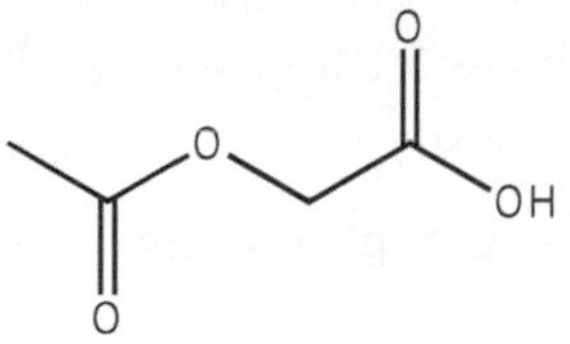	68
F	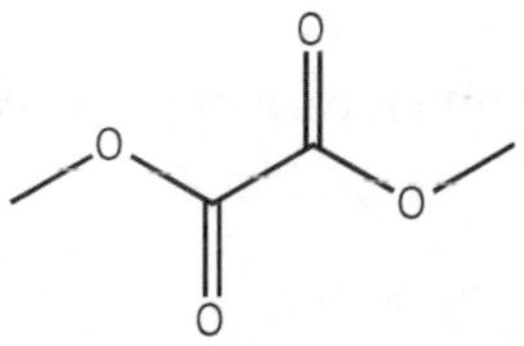	52

i What is the systematic name of compound **D**?

ii What is the molecular formula and molar mass of the isomers? *(2 marks)*

iii Predict the ^{1}H NMR spectra of **C** and **D**.

Include number of peaks, relative peak areas, types of proton, and splitting patterns. *(9 marks)*

iv A solution of compound **C** is reacted with an excess of sodium carbonate.

- Write an equation for the reaction using structural formulae for organic compounds.
- What type of reaction has taken place? *(2 marks)*

v Explain the difference in melting points. *(5 marks)*

c Compound **G** is a straight-chain dibasic organic acid. A student prepares a 250.0 cm^3 solution of 3.672 g of compound **G** in water.

The student titrates 25.0 cm^3 portions of this solution with 0.200 mol dm^{-3} NaOH(aq) in the burette.

The student's titration readings are shown below.

Titration	Trial	1	2	3
Final burette reading/cm^3	25.60	25.25	26.10	27.70
Initial burette reading/cm^3	0.00	0.50	1.00	2.50

i Outline how the student could prepare their 250.0 cm^3 solution, including apparatus and key procedures. *(3 marks)*

ii The maximum uncertainty in each burette reading is ±0.05 cm^3.

Calculate the percentage uncertainty in the titre for **Titration 1**. *(1 mark)*

iii Use the results of the student's analysis to identify compound **G**.

Show all your working. *(6 marks)*

4 A student carries out experiments to determine the formula of a hydrated Group 2 chloride, $\mathbf{M}Cl_2{\bullet}xH_2O$.

Experiment 1

The student first removes the water of crystallisation by heating a sample of the hydrated chloride in a crucible. The anhydrous chloride $\mathbf{M}Cl_2$ forms.

The results are shown below.

Mass of crucible /g	17.828
Mass of crucible + $\mathbf{M}Cl_2{\bullet}xH_2O$ /g	18.898
Mass of crucible + $\mathbf{M}Cl_2$ /g	18.464

Experiment 2

The student dissolves the anhydrous chloride $\mathbf{M}Cl_2$ in water and adds an excess of aqueous silver nitrate, $AgNO_3$(aq). A white precipitate forms which is filtered, washed with water, and dried. 1.150 g of the white precipitate are formed.

a Write ionic and full equations, with state symbols, for the formation of the white precipitate. *(2 marks)*

b Determine metal **M** and the formula of the anhydrous chloride, $\mathbf{M}Cl_2$. *(4 marks)*

c Determine the formula of the hydrated Group 2 chloride, $\mathbf{M}Cl_2{\bullet}xH_2O$. *(3 marks)*

d How could the student modify **Experiment 1** to be confident that all the water of crystallisation has been removed? *(1 mark)*

e In **Experiment 2**, explain how the results would be affected by the following errors.

i The precipitate had not been washed with water. *(2 marks)*

ii Less than an excess of aqueous silver nitrate had been added. *(2 marks)*

Answers to summary questions

18.1

1 a overall order = 3 *[1]*

b rate = $k[A]^2[B]$ *[1]*

2 a i Rate does not change. *[1]*

ii Rate increases × 9. *[1]*

iii Rate increases by × 64. *[1]*

b units = $dm^6\,mol^{-2}\,s^{-1}$ *[1]*

3 a rate = $k[A][B]^2$ $k = 12.5\,dm^6\,mol^{-2}\,s^{-1}$ *[3]*

b rate = $1.19 \times 10^{-4}\,mol\,dm^{-3}\,s^{-1}$ *[1]*

18.2

1 a *[1]*

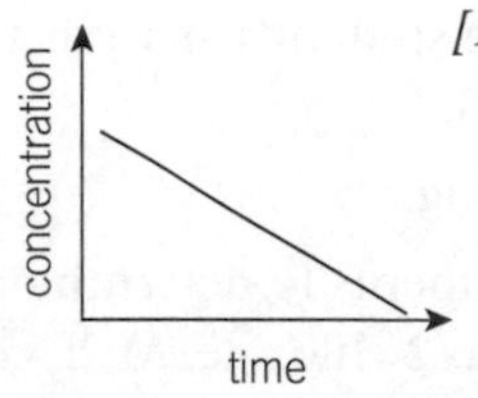

b *[1]*

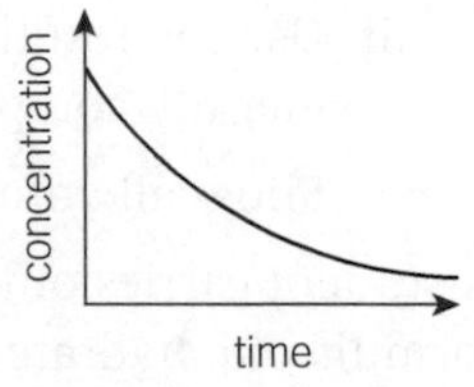

2 Reaction 1: $k = 0.0330\,s^{-1}$ *[1]*
Reaction 2: $k = 14.3\,s^{-1}$ *[1]*

3 a *[4]*

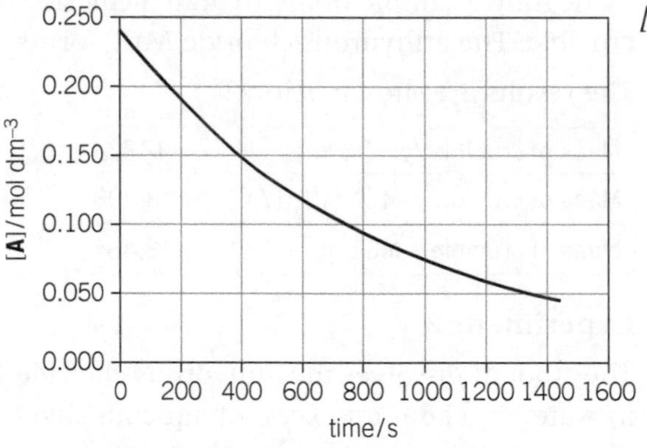

b i $\frac{0.24}{950} = 2.5 \times 10^{-4}\,mol\,dm^{-3}\,s^{-1}$ *[1]*

ii $\frac{0.20}{1400} = 1.4 \times 10^{-4}\,mol\,dm^{-3}\,s^{-1}$ *[1]*

c Two half-lives of 600 s which is a constant half-life *[2]*

d i $\frac{2.5 \times 10^{-4}}{0.240} = 1.04 \times 10^{-3}\,s^{-1}$ *[1]*

ii $\frac{\ln 2}{600} = 1.16 \times 10^{-3}\,s^{-1}$ *[1]*

18.3

1 a *[1]*

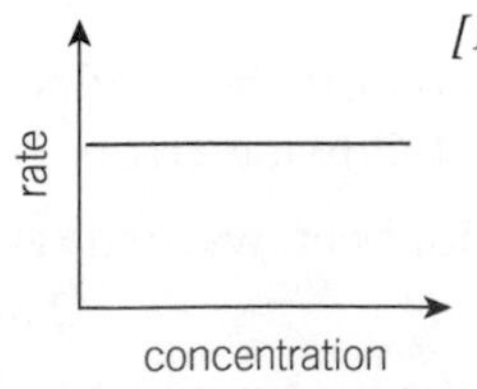

b *[1]*

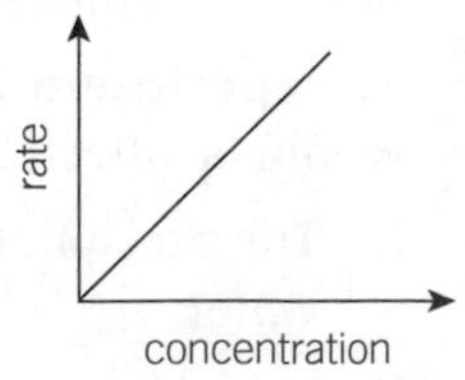

2 The rate constant is the gradient of the straight line *[1]*

3 a In $mol\,dm^3\,s^{-1}$:
0.0345; 0.0278; 0.0192; 0.0141; 0.00741 *[2]*

b *[4]*

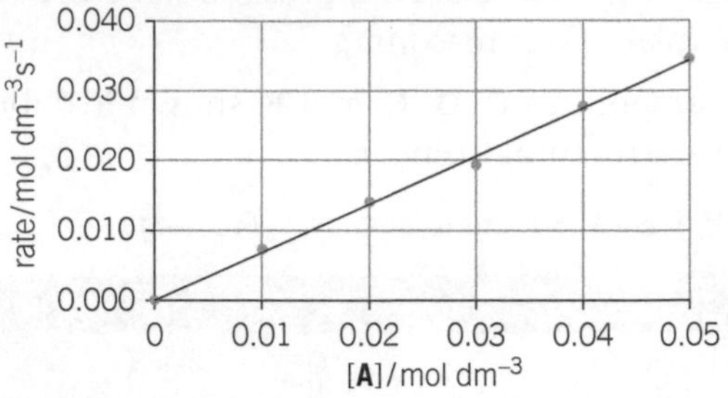

c i 1st order *[1]* ii $k = \frac{0.0345}{0.05} = 0.69\,s^{-1}$ *[2]*

18.4

1 a The slowest step in the reaction mechanism of a multi-step reaction. *[1]*

b The reactants in the rate-determining step give the species in the rate equation. *[1]*

2 a rate = $k[H_2(g)][ICl(g)]$ *[1]*

b $H_2(g) + 2ICl(g) \rightarrow 2HCl(g) + I_2(g)$ *[1]*

3 $2NO_2(g) \rightarrow NO(g) + NO_3(g)$ slow

$CO(g) + NO_3(g) \rightarrow CO_2(g) + NO_2(g)$ fast *[2]*

18.5

1 With increasing temperature, the rate constant and rate both increase. *[2]*

2 a $\ln k$ against $\frac{1}{T}$ *[1]*

b Gradient = $-\frac{E_a}{R}$; $\ln A$ = intercept on y axis. *[2]*

3 a *[4]*

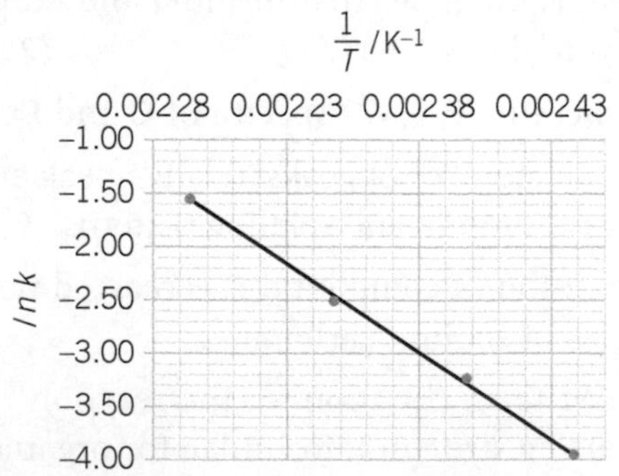

b Gradient = −16400 = $-E_a/RT$;
$E_a = \frac{16400}{1000} \times 8.314 = 136\,kJ\,mol^{-1}$ *[2]*

c $A = e^{36.0} = 4.31 \times 10^{15}$ *[1]*

19.1

1 a $K_c = \frac{[NO_2(g)]^2}{[N_2O_4(g)]}\,mol\,dm^{-3}$ *[2]*

b $K_c = \frac{[H_2O(g)]^2[Cl_2(g)]^2}{[HCl(g)]^4[O_2(g)]}\,dm^3\,mol^{-1}$ *[2]*

c $K_c = \frac{[H_2(g)]^4}{[H_2O(g)]^4}$ no units *[2]*

2 a $[N_2(g)] = 5 \times 3.25 \times 10^{-3} = 0.01625\,\text{mol dm}^{-3}$
$[H_2(g)] = 5 \times 6.25 \times 10^{-2} = 0.3125\,\text{mol dm}^{-3}$
$[NH_3(g)] = 5 \times 1.64 \times 10^{-3} = 8.20 \times 10^{-3}\,\text{mol dm}^{-3}$ *[3]*

b $K_c = \dfrac{[NH_3(g)]^2}{[N_2(g)][H_2(g)]^3} = \dfrac{(8.20 \times 10^{-3})^2}{0.01625 \times 0.3125^3}$
$= 0.136\,\text{dm}^6\,\text{mol}^{-2}$ *[3]*

3 a $n(CO) = 0.0250 - 4.50 \times 10^{-3} = 0.0205\,\text{mol}$;
$n(H_2) = 0.100 - 2 \times 4.50 \times 10^{-3} = 0.0910\,\text{mol}$ *[2]*

b $[CO(g)] = \dfrac{0.0250}{5.00} = 4.10 \times 10^{-3}\,\text{mol dm}^{-3}$;
$[H_2(g)] = \dfrac{0.0910}{5.00} = 0.0182\,\text{mol dm}^{-3}$;
$[CH_3OH(g)] = \dfrac{4.50 \times 10^{-3}}{5.00} = 9.00 \times 10^{-4}\,\text{mol dm}^{-3}$ *[3]*

c $K_c = \dfrac{[CH_3OH(g)]}{[CO(g)]\,[H_2(g)]^2} = \dfrac{9.00 \times 10^{-4}}{4.10 \times 10^{-3} \times 0.0182^2}$
$= 663\,\text{dm}^6\,\text{mol}^{-2}$ *[3]*

19.2

1 a Total number of gas moles = 15 + 16 + 9 = 40 mol
$x(Cl_2) = \dfrac{15}{40} = 0.375$; $x(O_2) = \dfrac{16}{40} = 0.400$;
$x(N_2) = \dfrac{9}{40} = 0.225$ *[3]*

b $p(N_2) = 0.375 \times 600 = 225\,\text{atm}$;
$p(H_2) = 0.400 \times 600 = 240\,\text{atm}$;
$p(NH_3) = 0.225 \times 600 = 135\,\text{atm}$ *[3]*

2 a $K_p = \dfrac{p(N_2O_4)}{p(NO_2)^2} = \dfrac{64}{150^2} = 0.00284\,\text{kPa}^{-1}$ *[3]*

b $K_p = \dfrac{p(C_2H_2) \times p(H_2)^3}{p(CH_4)^2} = \dfrac{2.52 \times 1.25^3}{14.42^2}$
$= 0.0237\,\text{atm}^2$ *[3]*

c $K_p = \dfrac{p(CO) \times p(H_2)}{p(H_2O)} = \dfrac{30 \times 85}{600} = 4.25\,\text{kPa}$ *[3]*

3 a $n(N_2) = 0.50 - \dfrac{0.80}{2} = 0.10\,\text{mol}$
$n(H_2) = 1.90 - 1.5 \times 0.80 = 0.70\,\text{mol}$ *[2]*

b Total number of gas moles = 0.10 + 0.70 + 0.80 = 1.60 mol
$x(N_2) = \dfrac{0.10}{1.60}$; $x(H_2) = \dfrac{0.70}{1.60}$; $x(NH_3) = \dfrac{0.80}{1.60}$ *[3]*
$p(N_2) = \dfrac{0.10}{1.60} \times 100 = 6.25\,\text{atm}$;
$p(H_2) = \dfrac{0.70}{1.60} \times 100 = 43.75\,\text{atm}$;

c $p(NH_3) = \dfrac{0.80}{1.60} \times 100 = 50.0\,\text{atm}$ *[3]*

$K_p = \dfrac{p(NH_3)^2}{p(N_2) \times p(H_2)^3} = \dfrac{50.0^2}{6.25 \times 43.75^3}$
$= 4.78 \times 10^{-3}\,\text{atm}^{-2}$ *[3]*

19.3

1 a K_c decreases. Equilibrium position shifts to left. *[2]*

b K_c increases. Equilibrium position shifts to right. *[2]*

2 a i K_c decreases *[1]*

ii K_c does not change *[1]*

iii K_c does not change *[1]*

iv K_c does not change. *[1]*

b Increasing temperature shifts equilibrium left as forward reaction is exothermic. *[1]*
Removing NH_3 shifts equilibrium right as concentration of ammonia decreases. *[1]*
Increasing pressure shifts equilibrium right as there are fewer gaseous moles of products. *[1]*
A catalyst does not affect the equilibrium position. *[1]*

3 a As temperature increases, K_c **decreases**. The system is no longer in equilibrium and $\dfrac{\text{products}}{\text{reactants}}$ is **greater** than K_c. The products **decrease** and the reactants **increase**, shifting the equilibrium position to the **left** until $\dfrac{\text{products}}{\text{reactants}}$ is equal to the new value of K_c. *[3]*

b As concentration of NH_3 decreases, the system is no longer in equilibrium and $\dfrac{\text{products}}{\text{reactants}}$ is **less** than K_c. The products **increase** and the reactants **decrease**, shifting the equilibrium position to the **right** until $\dfrac{\text{products}}{\text{reactants}}$ is once again equal to the value of K_c. *[3]*

20.1

1 a i A proton donor. *[1]* **ii** A proton acceptor. *[1]*

b i I^- *[1]* **ii** ClO_3^- *[1]* **iii** $CH_3CH_2CH_2COO^-$ *[1]*

2 a $CH_3COOH + OH^- \rightleftharpoons H_2O + CH_3COO^-$
Acid 1 Base 2 Acid 2 Base 1 *[1]*

b $HNO_2(aq) + CO_3^{2-}(aq) \rightleftharpoons NO_2^-(aq) + HCO_3^-(aq)$
Acid 1 Base 2 Base 1 Acid 2 *[1]*

c $NH_3(aq) + H_2SO_4(aq) \rightleftharpoons NH_4^+(aq) + HSO_4^-(aq)$
Base 2 Acid 1 Acid 2 Base 1 *[1]*

3 a $2HCl(aq) + Na_2CO_3(aq) \rightarrow 2NaCl(aq) + CO_2(g) + H_2O(l)$ *[1]*
$2H^+(aq) + CO_3^{2-}(aq) \rightarrow H_2O(l) + CO_2(g)$ *[1]*

b $Cu(OH)_2(s) + H_2SO_4(aq) \rightarrow CuSO_4(aq) + 2H_2O(l)$ *[1]*

$Cu(OH)_2(s) + 2H^+(aq) \rightarrow Cu^{2+}(aq) + 2H_2O(l)$ [1]

c $3Mg(s) + 2H_3PO_4(aq) \rightarrow Mg_3(PO_4)_2(aq) + 3H_2(g)$ [1]

$Mg(s) + 2H^+(aq) \rightarrow Mg^{2+}(aq) + H_2(g)$ [1]

20.2

1 a i pH = 3 [1] **ii** pH = 12 [1] **iii** pH = 7 [1]

b i 1×10^{-5} mol dm^{-3} [1] **ii** 1×10^{-8} mol dm^{-3} [1]

c i 10^6 times more [1] **ii** pH = 13 [1]

2 a i pH = 2.60 [1] **ii** pH = 5.09 [1]

iii pH = 7.43 [1]

b i 1.55×10^{-4} mol dm^{-3} [1]

ii 1.91×10^{-9} mol dm^{-3} [1]

iii 1.74×10^{-11} mol dm^{-3} [1]

3 a Concentration of diluted solution = 0.005 mol dm^{-3}. pH = –log 0.005 = 2.30 [2]

b $n(HCl) = \frac{0.73}{36.5} = 0.020$ mol.

$[HCl] = 0.020 \times \frac{1000}{250} = 0.080$ mol dm^{-3};

pH = –log 0.080 = 1.10 [3]

20.3

1 a $K_a = \frac{[H^+(aq)]\,[CH_3CH_2COO^-(aq)]}{[CH_3CH_2COOH(aq)]}$ [1]

b $K_a = \frac{[H^+(aq)]\,[NO_2^-(aq)]}{[HNO_2(aq)]}$ [1]

c $K_a = \frac{[H^+(aq)]\,[HSO_3^-(aq)]}{[H_2SO_3(aq)]}$ [1]

2 a i 2.86 [1] **ii** 6.50 [1] **iii** 9.24 [1]

b i 7.59×10^{-3} mol dm^{-3} [1]

ii 6.76×10^{-4} mol dm^{-3} [1]

iii 3.39×10^{-5} mol dm^{-3} [1]

3 a $HBrO(aq) \rightleftharpoons H^+(aq) + BrO^-(aq)$ [1]

$HCN(aq) \rightleftharpoons H^+(aq) + CN^-(aq)$ [1]

HBrO is the stronger acid as pK_a has the smaller number. [1]

b $ClCH_2COOH + C_6H_5COOH \rightleftharpoons$
Acid 1 Base 2

$C_6H_5COOH_2^+ + ClCH_2COO^-(aq)$
Acid 2 Base 1 [1]

20.4

1 a [H$^+$(aq)] from dissociation of water is negligible and [H$^+$(aq)] ~ [A$^-$(aq)]. [1]

b The small proportion of HA molecules that dissociates is negligible. [1]

2 a i $[H^+(aq)] = \sqrt{(6.46 \times 10^{-5} \times 1.00)}$
$= 8.04 \times 10^{-3}$ mol dm^{-3} [1]

pH = – log 8.04×10^{-3} = 2.09 [1]

ii $[H^+(aq)] = \sqrt{(6.46 \times 10^{-5} \times 0.250)}$
$= 4.02 \times 10^{-3}$ mol dm^{-3} [1]

pH = – log 4.02×10^{-3} = 2.40 [1]

iii $[H^+(aq)] = \sqrt{(6.46 \times 10^{-5} \times 4.20 \times 10^{-3})}$
$= 5.21 \times 10^{-4}$ mol dm^{-3} [1]

pH = – log 5.21×10^{-4} = 3.28 [1]

b $[H^+(aq)] = 10^{-2.32} = 4.79 \times 10^{-3}$ mol dm^{-3} [1]

$K_a = \frac{(4.79 \times 10^{-3})^2}{0.125} = 1.83 \times 10^{-4}$ mol dm^{-3} [1]

c $[H^+(aq)] = 10^{-3.64} = 2.29 \times 10^{-4}$ mol dm^{-3} [1]

$[HNO_2] = \frac{(2.29 \times 10^{-4})^2}{4.00 \times 10^{-4}} = 1.31 \times 10^{-4}$ mol dm^{-3} [1]

3 a H$^+$(aq) ~ [A$^-$(aq)] breaks down for very weak acids and very dilute solutions. [1]

$[HA]_{equilibrium}$ ~ $[HA]_{undissociated}$ breaks down for 'stronger' weak acids and concentrated solutions. [1]

b $K_a = 10^{-9.31} = 4.90 \times 10^{-10}$ mol dm^{-3} [1]

$[H^+(aq)] = \sqrt{(4.90 \times 10^{-10} \times 0.0250)}$
$= 3.50 \times 10^{-6}$ mol dm^{-3} [1]

$[CN^-(aq)] = [H^+(aq)] = 3.50 \times 10^{-6}$ mol dm^{-3} [1]

20.5

1 a $K_w = [H^+(aq)]\,[OH^-(aq)]$ [1]

b i $[H^+(aq)] = 10^{-4.00} = 1.00 \times 10^{-4}$ mol dm^{-3}

$[OH^-(aq)] = \frac{K_w}{[H^+(aq)]} = \frac{1.00 \times 10^{-14}}{1.00 \times 10^{-4}}$
$= 1.00 \times 10^{-10}$ mol dm^{-3} [2]

ii $[H^+(aq)] = 10^{-8.25} = 5.62 \times 10^{-9}$ mol dm^{-3}

$[OH^-(aq)] = \frac{K_w}{[H^+(aq)]} = \frac{1.00 \times 10^{-14}}{5.62 \times 10^{-9}}$
$= 1.78 \times 10^{-6}$ mol dm^{-3} [2]

2 a i $[H^+(aq)] = \frac{K_w}{[OH^-(aq)]} = \frac{1.00 \times 10^{-14}}{0.0191}$
$= 5.24 \times 10^{-13}$ mol dm^{-3}

pH = –log[H$^+$(aq)] = –log(5.24×10^{-13})
= 12.28 [2]

ii $[H^+(aq)] = \frac{K_w}{[OH^-(aq)]} = \frac{1.00 \times 10^{-14}}{4.19 \times 10^{-3}}$
$= 2.39 \times 10^{-12}$ mol dm^{-3}

pH = –log[H$^+$(aq)] = –log(2.39×10^{-12}) = 11.62 [2]

b i $[H^+(aq)] = 10^{-11.23} = 5.89 \times 10^{-12}$ mol dm^{-3}

$[OH^-(aq)] = \frac{K_w}{[OH^+(aq)]} = \frac{1.00 \times 10^{-14}}{5.89 \times 10^{-12}}$
$= 1.70 \times 10^{-3}$ mol dm^{-3} [2]

ii $[H^+(aq)] = 10^{-13.64} = 2.29 \times 10^{-14}$ mol dm^{-3}

$$[OH^-(aq)] = \frac{K_w}{[OH^+(aq)]} = \frac{1.00 \times 10^{-14}}{2.29 \times 10^{-14}}$$
$= 0.437\,mol\,dm^{-3}$ *[2]*

3 a $[OH^-(aq)] = 2 \times (2.49 \times 10^{-3})$
$= 4.98 \times 10^{-3}\,mol\,dm^{-3}$

$$[H^+(aq)] = \frac{K_w}{[OH^-(aq)]} = \frac{1.00 \times 10^{-14}}{4.98 \times 10^{-3}}$$
$= 2.01 \times 10^{-12}\,mol\,dm^{-3}$

$pH = -\log[H^+(aq)] = -\log(2.01 \times 10^{-12}) = 11.70$ *[3]*

b i $[H^+(aq)] = \sqrt{(5.48 \times 10^{-14})} = 2.34 \times 10^{-7}\,mol\,dm^{-3}$
$pH = -\log[H^+(aq)] = -\log(2.34 \times 10^{-7}) = 6.63$ *[2]*

ii $$[H^+(aq)] = \frac{K_w}{[OH^-(aq)]} = \frac{5.48 \times 10^{-14}}{0.0125}$$
$= 4.384 \times 10^{-12}\,mol\,dm^{-3}$

$pH = -\log[H^+(aq)] = -\log(4.384 \times 10^{-12})$
$= 11.36$ *[2]*

21.1

1 a Mix HCOOH(aq) with HCOONa(aq). *[1]*
Add NaOH(aq) to an excess of HCOOH(aq). *[1]*

b On addition of an acid, $H^+(aq)$ ions react with the conjugate base $HCOO^-(aq)$. *[1]*
The equilibrium position shifts to the left, removing most of the $H^+(aq)$ ions. *[1]*
On addition of an alkali, $OH^-(aq)$, the small concentration of $H^+(aq)$ ions reacts with the $OH^-(aq)$ ions:
$H^+(aq) + OH^-(aq) \rightarrow H_2O(l)$. *[1]*
HCOOH(aq) dissociates and the equilibrium position shifts to the right, restoring most of the $H^+(aq)$ ions. *[1]*

2 a $$[H^+(aq)] = 1.34 \times 10^{-5} \times \frac{0.200}{0.800}$$
$= 3.35 \times 10^{-6}\,mol\,dm^{-3}$ *[1]*
$pH = -\log(3.35 \times 10^{-6}) = 5.47$ *[1]*

b $$[H^+(aq)] = 6.28 \times 10^{-5} \times \frac{1.25}{0.25}$$
$= 3.14 \times 10^{-4}\,mol\,dm^{-3}$ *[1]*
$pH = -\log(3.14 \times 10^{-4}) = 3.50$ *[1]*

3 a $$[CH_3COOH] = 0.720 \times \frac{600}{1000} = 0.432\,mol\,dm^{-3}$$ *[1]*

$$[CH_3COO^-] = 0.300 \times \frac{400}{1000} = 0.120\,mol\,dm^{-3}$$ *[1]*

$$[H^+(aq)] = 1.74 \times 10^{-5} \times \frac{0.432}{0.120}$$
$= 6.264 \times 10^{-5}\,mol\,dm^{-3}$ *[1]*
$pH = -\log(6.264 \times 10^{-5}) = 4.20$ *[1]*

b $$n(CH_3COO^-) = 0.500 \times \frac{100}{1000} = 0.0500\,mol$$ *[1]*

$$n(CH_3COOH) = 0.200 \times \frac{400}{1000} - 0.0500$$
$= 0.0800 - 0.0500 = 0.0300\,mol$ *[1]*

$$[CH_3COOH] = 0.0300 \times \frac{1000}{500} = 0.0600\,mol\,dm^{-3}$$ *[1]*

$$[CH_3COO^-] = 0.0500 \times \frac{1000}{500} = 0.100\,mol\,dm^{-3}$$ *[1]*

$$[H^+(aq)] = 1.74 \times 10^{-5} \times \frac{0.0600}{0.100}$$
$= 1.044 \times 10^{-5}\,mol\,dm^{-3}$ *[1]*
$pH = -\log(1.044 \times 10^{-5}) = 4.98$ *[1]*

21.2

1 a $H_2CO_3(aq) \rightleftharpoons H^+(aq) + HCO_3^-(aq)$ *[1]*

$$K_a = \frac{[H^+(aq)]\,[HCO_3^-(aq)]}{[H_2CO_3(aq)]}$$ *[1]*

b On addition of an acid, $H^+(aq)$ ions react with HCO_3^- (aq). *[1]*
The equilibrium position shifts to the left, removing most of the $H^+(aq)$ ions. *[1]*
On addition of an alkali, the small concentration of $H^+(aq)$ ions reacts with the $OH^-(aq)$ ions:
$H^+(aq) + OH^-(aq) \rightarrow H_2O(l)$. *[1]*
$H_2CO_3(aq)$ dissociates and the equilibrium position shifts to the right, restoring most of the $H^+(aq)$ ions. *[1]*

2 $$K_a = \frac{[H^+(aq)]\,[HCO_3^-(aq)]}{[H_2CO_3(aq)]} = \frac{[H^+(aq)] \times 12}{1}$$ *[1]*

$$\therefore [H^+(aq)] = \frac{K_a \times 1}{12} = \frac{(7.9 \times 10^{-7}) \times 1}{12}$$
$= 6.58 \times 10^{-8}\,mol\,dm^{-3}$ *[1]*
$pH = -\log\,[H^+(aq)] = -\log\,(6.58 \times 10^{-8}) = 7.18$ *[1]*

3 For 7.35, $[H^+(aq)] = 10^{-pH} = 10^{-7.35}$
$= 4.47 \times 10^{-8}\,mol\,dm^{-3}$ *[1]*

$$\frac{[HCO_3^-(aq)]}{[H_2CO_3(aq)]} = \frac{K_a}{[H^+(aq)]} = \frac{7.9 \times 10^{-7}}{4.47 \times 10^{-8}} = \frac{17.7}{1}$$ *[1]*

For 7.45, $[H^+(aq)] = 10^{-pH} = 10^{-7.45}$
$= 3.55 \times 10^{-8}\,mol\,dm^{-3}$ *[1]*

$$\frac{[HCO_3^-(aq)]}{[H_2CO_3(aq)]} = \frac{7.9 \times 10^{-7}}{3.55 \times 10^{-8}} = \frac{22.3}{1}$$ *[1]*

21.3

1 a The vertical section shows when the pH changes rapidly on addition of a small volume of base. It coincides with exact neutralisation of an acid and base. *[2]*

b Different indicators have different pK_a values and different pH ranges. *[1]*

2 a i bromophenol blue, methyl orange *[1]*

ii phenolphthalein and metacresol purple. *[1]*

b There is no vertical section in the pH titration curve and the colour changes over addition of several cm^3 of solution. *[1]*

3 a Green *[1]*

b i $OH^-(aq)$ ions from the alkali react with $H^+(aq)$ ions in the equilibrium. *[1]*

The weak acid form of the indicator HA dissociates, shifting the equilibrium position to the right. *[1]*

The indicator colour changes to blue. *[1]*

ii $H^+(aq)$ ions from the acid react with the conjugate base $A^-(aq)$ form of the indicator. *[1]*

The equilibrium position shifts to the left. *[1]*

The indicator changes colour to yellow. *[1]*

22.1

1 a i Enthalpy change of atomisation of lithium. *[1]*

ii First electron affinity of bromine. *[1]*

iii Lattice enthalpy of calcium iodide. *[1]*

iv Second ionisation energy of magnesium. *[1]*

2 a i $K(s) + \frac{1}{2}Br_2(l) \rightarrow KBr(s)$ *[1]*

ii $2Na^+(g) + O^{2-}(g) \rightarrow Na_2O(s)$ *[1]*

iii $\frac{1}{2}F_2(g) \rightarrow F(g)$ *[1]*

b i

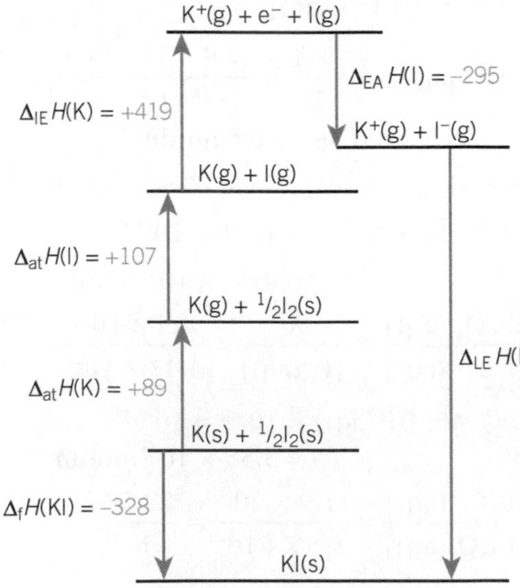

[5]

ii $89 + 107 + 419 + (-295) + \Delta_{LE}H^\ominus(KI) = -328$

$\therefore \Delta_{LE}H^\ominus(KI) = -328 - 320 = -648\,kJ\,mol^{-1}$ *[2]*

3 a $S^-(g) + e^- \rightarrow S^{2-}(g)$ *[1]*

b i

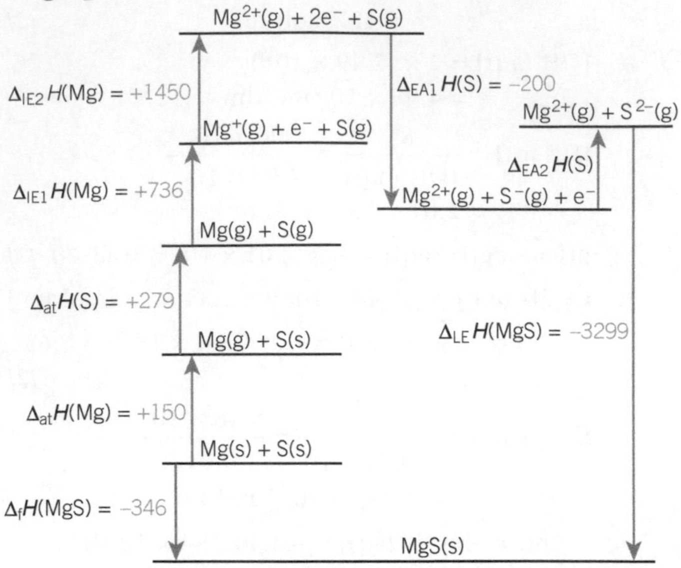

[5]

ii $+150 + 279 + 736 + 1450 + (-200) + \Delta_{EA2}H^\ominus(S) + (-3299) = -346$

$\Delta_{EA2}H^\ominus(S) = -346 + 884 = +538\,kJ\,mol^{-1}$ *[2]*

22.2

1 a Enthalpy change of hydration $\Delta_{hyd}H$ is the enthalpy change when one mole of gaseous ions dissolve in water to form aqueous ions. *[1]*

b i Enthalpy change of hydration of F^-. *[1]*

ii Enthalpy change of solution of LiF. *[1]*

2 a i $Ca^{2+}(g) + aq \rightarrow Ca^{2+}(aq)$ *[1]*

ii $Li^+(g) + Br^-(g) \rightarrow LiBr(s)$ *[1]*

iii $CaCl_2(s) \rightarrow Ca^{2+}(aq) + 2Cl^-(aq)$ *[1]*

b $\Delta_{LE}H^\ominus(NaCl) + 4 = (-406) + (-378) = -784$

$\Delta_{LE}H^\ominus(NaCl) = -784 - 4 = -788\,kJ\,mol^{-1}$ *[2]*

3 a

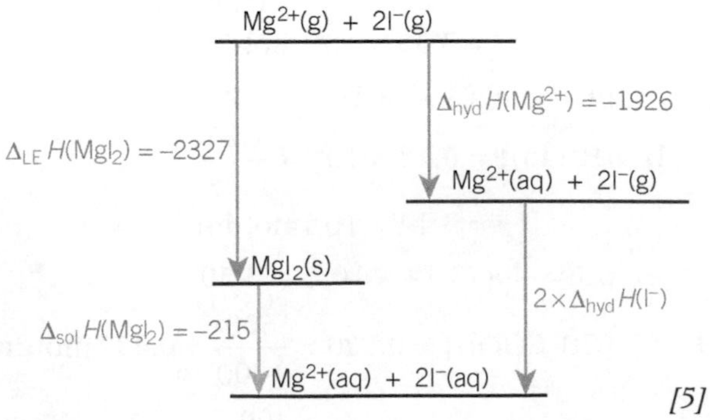

[5]

ii $-2327 + (-215) = -1926 + (2 \times \Delta_{hyd}H(I^-))$

$\Delta_{hyd}H(I^-) = (-2542 + 1926)/2 = -308\,kJ\,mol^{-1}$ *[2]*

22.3

1 As ionic size decreases and ionic charge increases:

- attraction between ion and water molecules increases
- enthalpy change of hydration becomes more exothermic. *[2]*

2 a The ionic size of K^+ is greater than ionic size of Na^+. The larger K^+ ions attract Br^- ions less strongly than Na^+ ions. Therefore $\Delta_{LE}H$(KBr) is less exothermic than $\Delta_{LE}H$(NaBr). *[2]*

b Ca^{2+} has a greater ionic charge than Na^+. The greater charge in Ca^{2+} attracts water molecules more strongly than Na^+ ions. Therefore $\Delta_{hyd}H(Ca^{2+})$ is more exothermic than $\Delta_{hyd}H(Na^+)$. *[2]*

3 Na^+ and Ca^{2+} have similar ionic radii and any difference in lattice enthalpy must be the result of different charges. Mg^{2+} ions have a smaller ionic size than Na^+. If Na^+ and Mg^{2+} are compared, it is impossible to know whether lattice enthalpy is affected by ionic size, or ionic charge, or both. *[2]*

22.4

1 a Entropy increases as liquid → gas. A gas has more entropy. *[1]*

b Entropy increases as more moles of gas in products (0 mol → 1 mol). *[1]*

c Entropy decreases as fewer moles of gas in products (3 mol → 1 mol). *[1]*

2 a $\Delta S = [(2 \times 27.3) + (3 \times 213.6)] - [87.4 + (3 \times 197.7)] = +14.9\,J\,K^{-1}\,mol^{-1}$ *[2]*

b $\Delta S = [4 \times 213.6 + (5 \times 69.9)] - [310.1 + (6.5 \times 205.0)] = -438.7\,J\,K^{-1}\,mol^{-1}$ *[2]*

3 a $-360.7 = [(8 \times 213.6) + (9 \times 69.9)] - [S^\ominus(C_8H_{18}(g)) + (12.5 \times 205.0)]$

$S^\ominus(C_8H_{18}(g)) = 2337.9 - 2562.5 + 360.7 = +136.1\,J\,K^{-1}\,mol^{-1}$ *[2]*

b $-198.8 = [(2 \times 192.3)] - [191.6 + (3 \times S^\ominus(H_2(g))]$

$3 \times S^\ominus(H_2(g)) = 384.6 - 191.6 + 198.8 = 391.8$

$S^\ominus(H_2(g)) = 391.8/3 = +130.6\,J\,K^{-1}\,mol^{-1}$ *[2]*

22.5

1 a $\Delta G < 0$ or $\Delta H - T\Delta S < 0$ *[1]*

b Although ΔG may be negative, there may be a large activation energy resulting in a very slow rate. *[1]*

2 a More moles of gas in products (0 mol → 1 mol). *[1]*

b $\Delta S = 160.8/1000 = 0.1608\,kJ\,mol^{-1}\,K^{-1}$

$\Delta G = 224.5 - 298 \times 0.1608 = +176.6\,kJ\,mol^{-1}$ *[1]*

As $\Delta G > 0$, the reaction is not feasible at 25 °C. *[1]*

c For minimum temperature, $\Delta G = 0$;

$\Delta H - T\Delta S = 0$: $T = \dfrac{\Delta H}{\Delta S}$ *[1]*

Minimum temperature $T = \dfrac{224.5}{0.1608} = 1396\,K = 1123\,°C$ *[1]*

3 a $\Delta S = [68.7 + (2 \times 240) + 0.5 \times 205.0] - 213 = +438.2\,J\,mol^{-1}\,K^{-1}$ *[1]*

$\Delta H = [-217.3 + (2 \times 33.2)] - (-451.9) = +301.0\,kJ\,mol^{-1}$ *[1]*

b $\Delta S = 438.2/1000 = 0.4382\,kJ\,mol^{-1}\,K^{-1}$

$\Delta G = 301.0 - 298 \times 0.4382 = +170.4\,kJ\,mol^{-1}$ *[1]*

As $\Delta G > 0$, the reaction is not feasible at 25 °C. *[1]*

c For minimum temperature $\Delta G = 0$;

$\Delta H - T\Delta S = 0$: $T = \dfrac{\Delta H}{\Delta S}$ *[1]*

Minimum temperature $T = \dfrac{301.0}{0.4382} = 687\,K = 414\,°C$ *[1]*

23.1

1 a Al 0 to +3; Cu +2 to 0. Al reducing agent, $CuSO_4$ oxidising agent. *[3]*

b Br −1 to 0; S +6 to +4. HBr reducing agent, H_2SO_4 oxidising agent. *[3]*

2 a $6I^- + Cr_2O_7^{2-} + 14H^+ \rightarrow 3I_2 + 2Cr^{3+} + 7H_2O$ *[2]*

b $2MnO_4^- + 6H^+ + 5H_2S \rightarrow 2Mn^{2+} + 5S + 8H_2O$ *[2]*

3 a i $MnO_4^- + 8H^+ + 5Cl^- \rightarrow Mn^{2+} + 4H_2O + 2\frac{1}{2}Cl_2$ *[2]*

ii $2VO_3^- + SO_2 + 4H^+ \rightarrow 2VO^{2+} + SO_4^{2-} + 2H_2O$ *[2]*

b $Sn + 4HNO_3 \rightarrow SnO_2 + 4NO_2 + 2H_2O$ *[2]*

23.2

1 a The titration does not need an indicator. At the end point, the colour changes from colourless to the first permanent pink colour. *[1]*

b Sulfuric acid provides H^+ ions for the manganate(VII) reduction (H^+ is in the equation). *[1]*

2 a Mn is reduced from +7 to +2; Fe is oxidised from +2 to +3. *[2]*

b $n(MnO_4^-) = 0.0250 \times \dfrac{24.80}{1000} = 6.20 \times 10^{-4}\,mol$ *[1]*

$(Fe^{2+}) = 5 \times 6.20 \times 10^{-4} = 3.10 \times 10^{-3}\,mol$ *[1]*

Concentration of $FeSO_4 = 3.10 \times 10^{-3} \times \dfrac{1000}{25.0} = 0.124\,mol\,dm^{-3}$ *[1]*

3 $n(MnO_4^-) = 0.0200 \times \dfrac{21.80}{1000} = 4.36 \times 10^{-4}\,mol$ *[1]*

$n(Fe^{2+}) = \mathbf{5} \times 4.36 \times 10^{-4} = 2.18 \times 10^{-3}\,mol$ *[1]*

$n(Fe^{2+})$ in 250 cm³ solution $= \mathbf{10} \times 2.18 \times 10^{-3} = 2.18 \times 10^{-2}\,mol$ *[1]*

Mass of $FeSO_4 \bullet 7H_2O$
$= n \times M = 2.18 \times 10^{-2} \times 277.9 = 6.05822\,g$ [1]

Percentage purity $= \frac{6.05822}{7.18} \times 100 = 84.4\ \%$ (to 3 s.f.) [1]

23.3

1 a Thiosulfate is added until the colour fades to pale straw. Starch is then added, which changes colour from blue-black to colourless at the end point. [2]

b KI provides I^- ions to reduce all of the oxidising agent. The I^- ions are oxidised to iodine for the titration. [2]

2 a I is reduced from 0 to –1; S is oxidised from +2 to +2.5. [2]

b $n(S_2O_3^{2-}) = 0.0150 \times \frac{22.40}{1000} = 3.36 \times 10^{-4}\,mol$ [1]

$n(Cu^{2+}) = n(S_2O_3^{2-}) = 3.36 \times 10^{-4}\,mol$ [1]

Concentration of $CuSO_4 = 3.36 \times 10^{-4} \times \frac{1000}{25.0}$

$= 0.01344\,mol\,dm^{-3}$ [1]

3 $n(S_2O_3^{2-}) = 0.0240 \times \frac{25.50}{1000} = 6.12 \times 10^{-4}\,mol$ [1]

$n(Cu^{2+}) = n(S_2O_3^{2-}) = 6.12 \times 10^{-4}\,mol$ [1]

$n(Cu^{2+})$ in $250\,cm^3$ solution $= \mathbf{10} \times 6.12 \times 10^{-4}$
$= 6.12 \times 10^{-3}\,mol$ [1]

Molar mass, M, of $CuCl_2 \bullet xH_2O = \frac{m}{n} = \frac{1.043}{6.12 \times 10^{-3}}$
$= 170.4\,g\,mol^{-1}$ [1]

M of $CuCl_2 = 134.5$; $x = \frac{170.4 - 134.5}{18.0} = \frac{35.9}{18.0} = 2$

Formula = $CuCl_2 \bullet 2H_2O$ [1]

23.4

1 a The standard electrode potential is the e.m.f. of a half-cell connected to a standard hydrogen half-cell under standard conditions of 298 K, solution concentrations of $1\,mol\,dm^{-3}$, and a pressure of 100 kPa. [3]

b i electrons [1]

ii ions [1]

2 a The + electrode has the more positive $E^\ominus$ value. The – electrode has the more negative $E^\ominus$ value. [1]

b i 0.78 V [1]

ii 2.46 V [1]

iii 3.14 V [1]

3 a reduction: $Cu^{2+} + 2e^- \rightarrow Cu$ [1]

oxidation: $Fe \rightarrow Fe^{2+} + 2e^-$ [1]

overall: $Cu^{2+} + Fe \rightarrow Cu + Fe^{2+}$ [1]

b reduction: $Ag^+ + e^- \rightarrow Ag$ [1]

oxidation: $Al \rightarrow Al^{3+} + 3e^-$ [1]

overall: $3Ag^+ + Al \rightarrow 3Ag + Al^{3+}$ [1]

c reduction: $Fe^{3+} + e^- \rightarrow Fe^{2+}$ [1]

oxidation: $Mg \rightarrow Mg^{2+} + 2e^-$ [1]

overall: $2Fe^{3+} + Mg \rightarrow 2Fe^{2+} + Mg^{2+}$ [1]

23.5

1 a List the $E^\ominus$ values sorted with most negative at the top. The strongest oxidising agent is at the bottom on the left. The strongest reducing agent is at the top on the right. [2]

b There may be a large activation energy, slowing down the reaction rate. [1]

The concentration, temperature, and pressure may not be standard. [1]

The reactants may not be aqueous. [1]

2 a i $I_2(aq)$ [1] ii $Mg(s)$ [1]

b i $Mg(s)$ [1] ii $Cr^{3+}(aq)$ and $I_2(aq)$ [1]

iii $Mg(s) + 2Cr^{3+}(aq) \rightarrow Mg^{2+}(aq) + 2Cr^{2+}(aq)$

$Mg(s) + I_2(aq) \rightarrow Mg^{2+}(aq) + 2I^-(aq)$ [2]

3 a $2Ag^+(aq) + I^-(aq) + 2OH^-(aq) \rightarrow 2Ag(s) + IO^-(aq) + H_2O(l)$ [1]

$3Ag^+(aq) + Cr(OH)_3(s) + 5OH^-(aq) \rightarrow 3Ag(s) + CrO_4^{2-}(aq) + 4H_2O(l)$ [1]

$3IO^-(aq) + 2Cr(OH)_3 + 4OH^-(aq) \rightarrow 3I^-(aq) + 2CrO_4^{2-}(aq) + 5H_2O(l)$ [1]

23.6

1 a A primary cell is just used once and cannot be recharged. A secondary cell can be recharged. [1]

b A fuel cell uses the energy from the reaction of a fuel with oxygen to create a voltage. [1]

2 a $2NiOOH + H_2 \rightarrow 2Ni(OH)_2$ [1]

b $E^\ominus = +0.52\,V$ [1]

3 a $E^\ominus = -2.35\,V$ [1]

b The O_2 half-cell because E is more positive than Al half-cell. [1]

c $Al \rightarrow Al^{3+} + 3e^-$ [2]

24.1

1 They form compounds in which the transition element has different oxidation states.

They form coloured compounds.

The elements and their compounds can act as catalysts. [3]

2 a i $1s^2 2s^2 2p^6 3s^2 3p^6 3d^8 4s^2$ [1]

ii $1s^2 2s^2 2p^6 3s^2 3p^6 3d^4$ [1]

b i +3 [1] ii +6 [1] iii +6 [1] iv +5 [1]

3 Scandium and zinc are d-block metals as their highest energy electrons occupy the 3d sub-shell:

Sc $1s^22s^22p^63s^23p^63d^14s^2$; Zn $1s^22s^22p^63s^23p^63d^{10}4s^2$

Scandium only forms the Sc^{3+} ion which has the electron configuration of $1s^22s^22p^63s^23p^6$.

Zinc only forms the Zn^{2+} ion which has the electron configuration of $1s^22s^22p^63s^23p^63d^{10}$. A transition element forms an ion with an incomplete d sub-shell. Sc^{3+} has an empty d sub-shell. Zn^{2+} has a full d sub-shell. *[4]*

24.2

1 a i A ligand is as a molecule or ion that donates a pair of electrons to a central metal ion to form a coordinate or dative covalent bond. *[1]*

ii A bidentate ligand is a molecule or ion that donates two pairs of electrons to a central metal ion to form two coordinate or dative covalent bonds. *[1]*

iii The coordination number is the number of coordinate bonds attached to the central metal ion. *[1]*

b 6, 90° *[2]*

2 a i Octahedral, 90°, 6 *[3]*

ii tetrahedral, 109.5°, 4. *[3]*

b i +2 *[1]* **ii** +3 *[1]*

c i $[CuF_6]^{4-}$ *[1]* **ii** $Fe(CN)_6]^{3-}$ *[1]*

iii $[Co(NH_3)_4Cl_2]^+$ *[1]*

iv $[Ni(H_2NCH_2CH_2NH_2)_3]^{2+}$ *[1]*

3 a $[Fe((COO)_2)_3]^{3-}$ *[1]* **b** $[Ni(NH_3)_2Cl_2]$ *[1]*

24.3

1 *cis*-platin is an anti-cancer drug and it acts by binding to DNA preventing cell division. *[2]*

2 a *[2]*

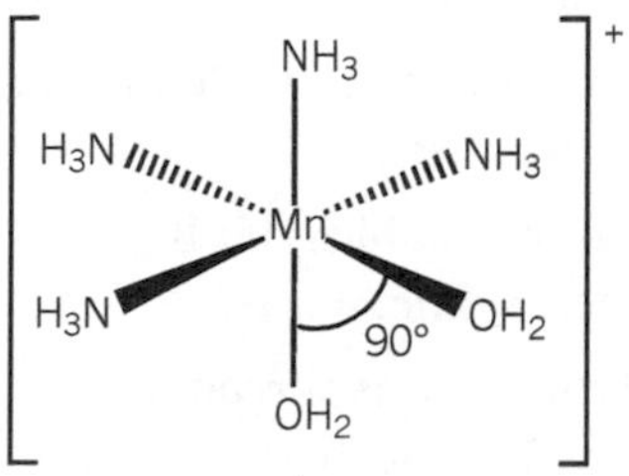

cis

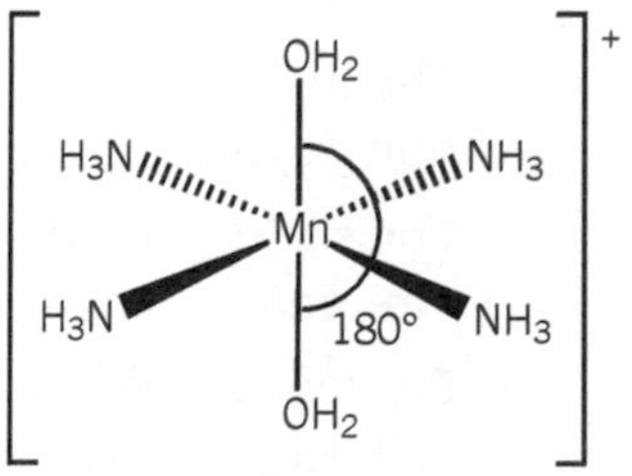

trans

b *[2]*

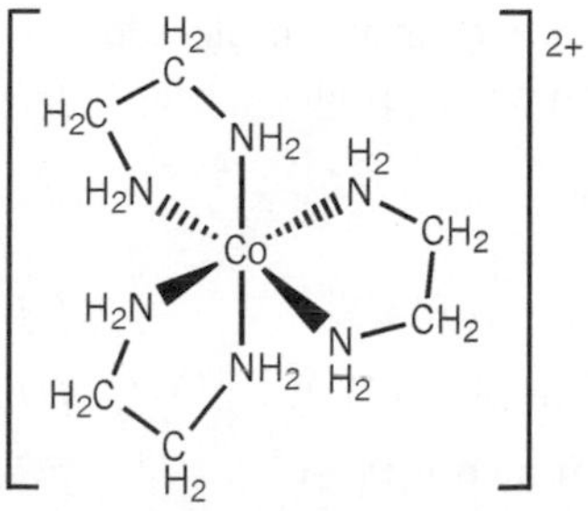

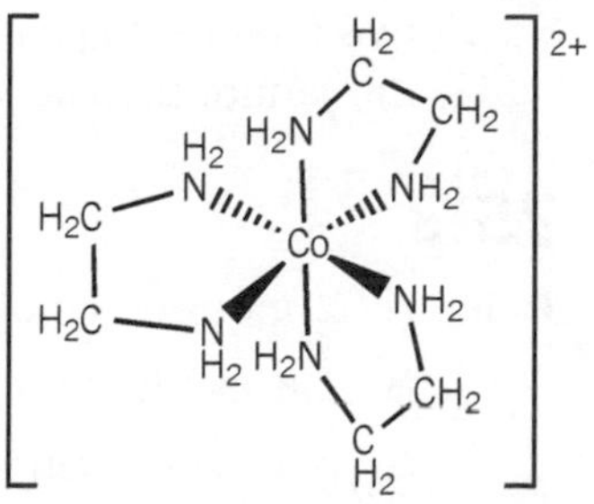

3 a $[Mn((COO)_2)_2(H_2O)_2]^{2-}$ *[1]*

b

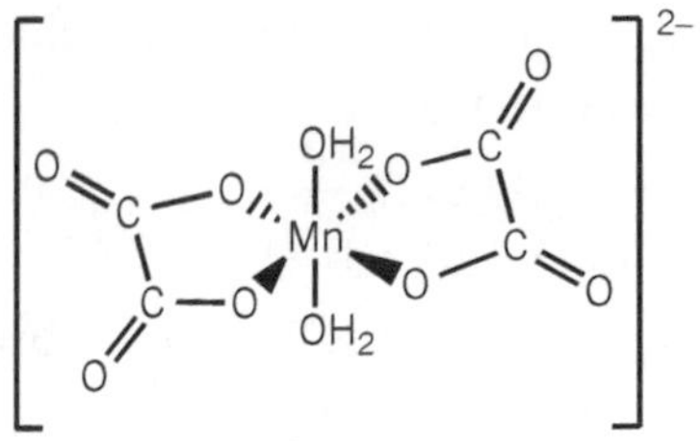

trans

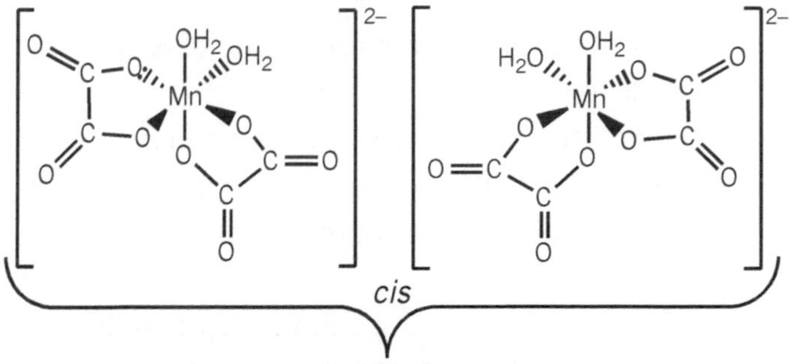

cis

Optical *[3]*

24.4

1 a Ligand substitution is a reaction of a complex ion in which one ligand is replaced by another ligand. *[1]*

b i violet *[1]* **ii** purple *[1]*

2 a $Mn^{2+}(aq) + 2OH^-(aq) \rightarrow Mn(OH)_2(s)$; light-brown precipitate *[2]*

b $Cu^{2+}(aq) + 2OH^-(aq) \rightarrow Cu(OH)_2(s)$; blue precipitate *[2]*

c $Fe^{3+}(aq) + 3OH^-(aq) \rightarrow Fe(OH)_3(s)$ orange-brown precipitate *[2]*

3 a A: pale blue, $[Cu(H_2O)_6]^{2+}$ *[2]*

B: yellow, $[CuCl_4]^{2-}$ *[2]*

C: pale blue precipitate, $Cu(OH)_2$ *[2]*

D: deep blue $[Cu(NH_3)_4(H_2O)_2]^{2+}$ *[2]*

b Ligand substitution *[1]*

c i $Cu(H_2O)_6]^{2+} + 4Cl^- \rightarrow [CuCl_4]^{2-} + 6H_2O$ *[1]*

ii $Cu(H_2O)_6]^{2+} + 4NH_3 \rightarrow [Cu(NH_3)_4(H_2O)_2]^{2+} + 4H_2O$ *[1]*

24.5

1 a i pale-green *[1]*

ii pale-yellow *[1]*

iii orange *[1]*

iv yellow *[1]*

2 a i H^+/MnO_4^-
Fe from +2 to +3; Mn from +7 to +2 *[3]*

ii OH^-/H_2O_2
Cr from +3 to +6; O from –1 to –2 *[3]*

b i I^-; Fe from +3 to +2; I from –1 to 0 *[3]*

ii Zn; Cr from +6 to +3; Zn from 0 to +2 *[3]*

iii I^-; Cu from +2 to +1; I^- from –1 to 0 *[3]*

3 a Cu_2O is heated with dilute sulfuric acid. *[1]*

Cu^+, have been simultaneously reduced to Cu and oxidised to Cu^{2+}. *[1]*

$Cu_2O(s) + H_2SO_4(aq) \rightarrow$
$Cu(s) + CuSO_4(aq) + H_2O(l)$ *[1]*

Cu_2O is a red-brown solid, Cu is a brown solid, $CuSO_4(aq)$ is a blue solution. *[1]*

b i Add $AgNO_3(aq)$. NaCl gives a white precipitate of AgCl, soluble in dilute $NH_3(aq)$.

NaBr gives a cream precipitate of AgBr, insoluble in dilute $NH_3(aq)$. *[3]*

ii Add NaOH(aq). $FeSO_4$ gives a green precipitate of $Fe(OH)_2$.

$Fe_2(SO_4)_3$ gives an orange-brown precipitate of $Fe(OH)_3$. *[3]*

iii Add NaOH(aq) and warm. NH_4Cl produces NH_3 gas which turns indicator blue.

$CrCl_3$ gives a green precipitate of $Cr(OH)_3$, dissolving in excess NaOH to form a green solution of $[Cr(OH)_6]^{3-}$. *[3]*

25.1

1 a Spread out. *[1]*

b i All carbon–carbon bond lengths are the same. In the Kekulé model, the C=C bond lengths would be shorter than the C–C bond lengths. *[2]*

ii Kekulé benzene should react in a similar way to alkenes with bromine and by electrophilic addition. Actual benzene does not react with bromine and does not react by electrophilic addition. *[2]*

iii Kekulé benzene would have an enthalpy change of hydrogenation of $-360\,kJ\,mol^{-1}$ equivalent to 3 C=C bonds. The actual enthalpy change is only $-208\,kJ\,mol^{-1}$ showing that benzene is more stable than Kekulé benzene. *[2]*

2 Each carbon atom has one electron in a p-orbital at right angles to the plane of the σ-bonded carbon and hydrogen atoms. Adjacent p-orbital electrons overlap sideways above and below the plane of the carbon atoms to form a ring of electron density. The overlapping orbitals create a system of pi-bonds which is delocalised (spread out) around all six C atoms in the ring structure. *[3]*

3 a

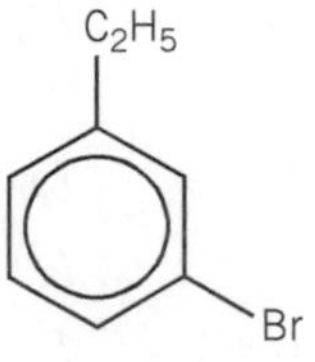

[1]

b

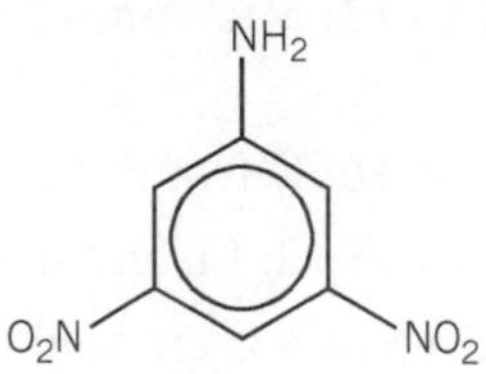

[1]

c

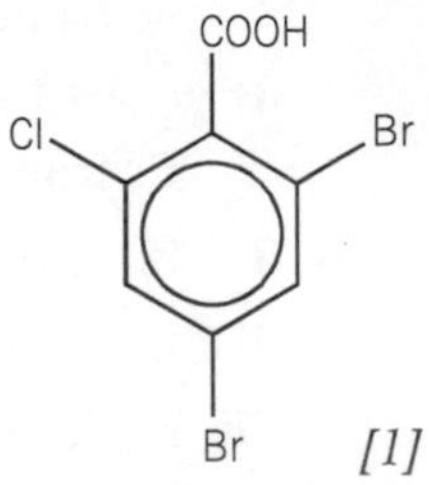

[1]

4 a 2-methylnitrobenzene *[1]*

b 1,2,3-trimethylbenzene *[1]*

c 2-bromo-5-methylphenol *[1]*

25.2

1 a Electrophilic substitution. *[1]*

b i Concentrated H_2SO_4 *[1]*

$C_6H_6 + HNO_3 \rightarrow C_6H_5NO_2 + H_2O$ *[1]*

ii Iron bromide *[1]*

$C_6H_6 + Br_2 \rightarrow C_6H_5Br + HBr$ *[1]*

2 a 4-nitromethylbenzene/4-methylnitrobenzene *[1]*

b $HNO_3 + H_2SO_4 \rightarrow NO_2^+ + HSO_4^- + H_2O$ *[1]*

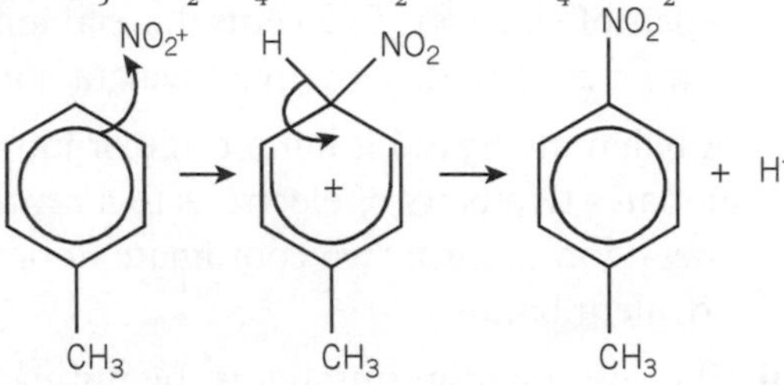

[2]

$H^+ + HSO_4^- \rightarrow H_2SO_4$ *[1]*

3 a CH_3COCl and $AlCl_3$ catalyst *[2]*

b $CH_3COCl + AlCl_3 \rightarrow CH_3CO^+ + AlCl_4^-$ *[1]*

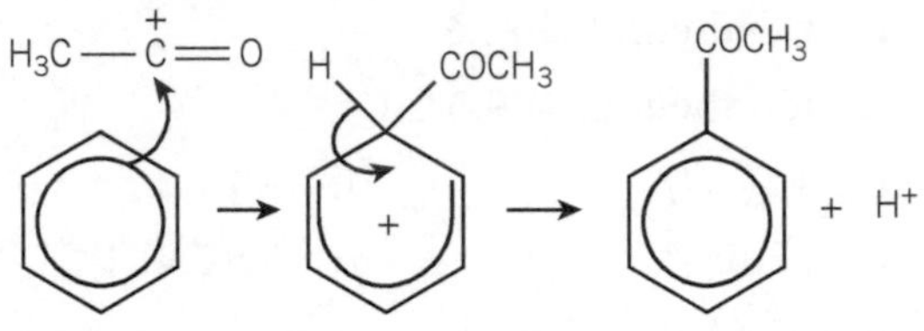

[2]

$H^+ + AlCl_4^- \rightarrow AlCl_3 + HCl$ *[1]*

4 a i Electrophilic addition *[1]*

ii Electrophilic substitution *[1]*

b i $C_3H_6 + Br_2 \rightarrow C_3H_6Br_2$ *[1]*

ii $C_6H_6 + Br_2 \rightarrow C_6H_5Br + HBr$ *[1]*

c The π-bond in the alkene has localised electrons above and below the C=C bond. This produces an area of high electron density.

The high electron density induces a dipole in the non-polar Br_2 molecule: $^{\delta+}Br\text{–}Br^{\delta-}$.

The slightly positive Br atom enables the Br_2 molecule to act as an electrophile.

Benzene has delocalised π-electrons spread above and below the carbon ring. The electron density is less than in a C=C double bond in an alkene. There is insufficient π-electron density around any two C atoms to polarise the bromine molecule. Br_2 cannot then act as an electrophile. *[3]*

25.3

1 a A 2-methylphenol (a phenol) *[1]*

B 2,6-dichlorophenol (a phenol) *[1]*

C phenylmethanol (an alcohol) *[1]*

b i Indicator turns red *[1]* **ii** No change *[1]*

2 a $C_6H_5OH \rightleftharpoons C_6H_5O^- + H^+$ *[1]*

b $C_6H_5OH + NaOH \rightarrow C_6H_5ONa + H_2O$ *[1]*

c Phenol reacts with bromine water at room temperature. *[1]*

Benzene reacts with bromine and a halogen carrier, e.g. $FeBr_3$. *[1]*

Phenol reacts with dilute nitric acid at room temperature. *[1]*

Benzene reacts with concentrated nitric and sulfuric acids at 50 °C. *[1]*

3 a A lone pair from the oxygen p-orbital of the –OH group is donated into the π-system of phenol. *[1]*

The electron density of the aromatic ring in phenol increases. *[1]*

The increased electron density attracts electrophiles more strongly than with benzene. *[1]*

b B is the only compound that effervesces with Na_2CO_3. *[1]*

Compound A turns indicator red **and** decolourises bromine. *[1]*

Compound D decolourises bromine but does not turn indicator red. *[1]*

Compound C does not react in any of the tests. *[1]*

25.4

1 a 2,4- *[1]* **b** 3- *[1]* **c** 3- *[1]* **d** 2,4- *[1]*

2 a *[1]* **b** *[1]* **c** *[1]*

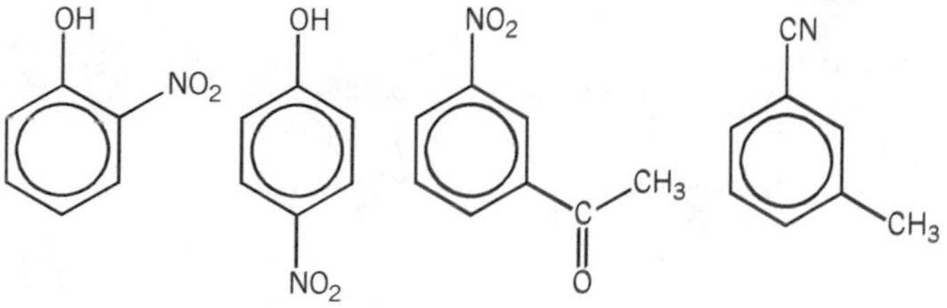

3 a *[2]*

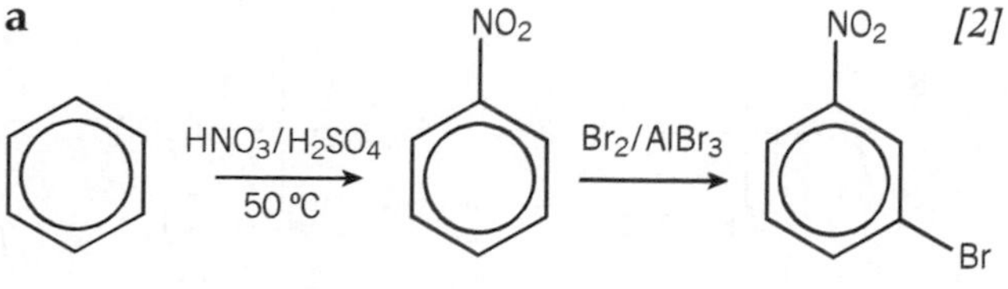

b *[2]*

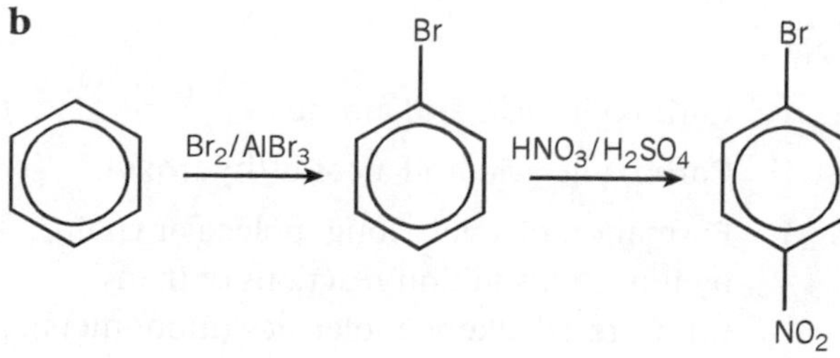

26.1

1 a In aldehydes the carbonyl functional group is at the end of a carbon chain. *[1]*

In ketones, the carbonyl functional group is between two carbon atoms in the carbon chain. *[1]*

b A species that donates an electron pair to a carbon atom to form a dative covalent bond *[1]*

c Nucleophilic addition *[1]*

2 a i $K_2Cr_2O_7/H_2SO_4$ *[1]*

$CH_3CHO + [O] \rightarrow CH_3COOH$ *[1]*

ii $NaCN/H_2SO_4$ *[1]*

$CH_3CHO + HCN \rightarrow CH_3CH(OH)CN$ *[1]*

b i $NaBH_4$ *[1]*

c i $CH_3CH_2CH_2CH_2CHO + 2[H] \rightarrow CH_3CH_2CH_2CH_2CH_2OH$ *[1]*

ii $CH_3CH_2COCH_3 + 2[H] \rightarrow CH_3CH_2CHOHCH_3$ *[1]*

3

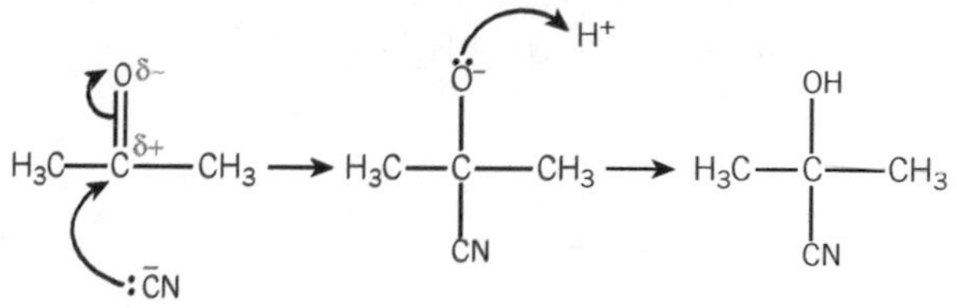

[4]

26.2

1 a A yellow/orange precipitate confirms a carbonyl group in aldehydes and ketones. *[2]*

b A silver mirror confirms an aldehyde. *[2]*

2 a $RCHO + [O] \rightarrow RCOOH$ *[1]*

$Ag^+(aq) + e^- \rightarrow Ag(s)$ *[1]*

b A compound is added to 2,4-DNP solution. *[1]*

The yellow/orange precipitate is filtered, recystallised and dried. *[1]*

The melting point is measured and compared with known melting points of 2,4-DNP derivatives. *[1]*

3 Add compounds to Tollens' reagent and warm. CH_3CH_2CHO and $CH_3CH_2CH_2CHO$ are aldehydes and would produce a silver mirror. *[1]*

Add compounds to 2,4-DNP solution. CH_3CH_2CHO, $(CH_3)_2CHCOCH_3$, and $CH_3CH_2CH_2CHO$ are carbonyl compounds and would produce a yellow/orange precipitate. *[1]*

The compound that does not form a silver mirror but does form a yellow/orange precipitate is the ketone, $(CH_3)_2CHCOCH_3$. *[1]*

The compound which gives no observation with Tollens' reagent or 2,4-DNP is the alcohol, $CH_3CH_2CH_2CH_2OH$. *[1]*

The 2,4 DNP derivatives for the two aldehydes are filtered, recrystallised, and dried. The melting points are measured and compared with known melting points of 2,4-DNP derivatives. *[2]*

26.3

1 a The carboxyl group contains polar C=O and O–H bonds which can form hydrogen bonds with water. *[2]*

b A weak acid partially dissociates, e.g. $HCOOH(aq) \rightleftharpoons H^+(aq) + HCOO^-(aq)$ *[2]*

2 a $2CH_3CH_2COOH(aq) + CaO(s) \rightarrow (CH_3CH_2COO^-)_2Ca^{2+}(aq) + H_2O(l)$ *[1]*

b $2HCOOH(aq) + CaCO_3(s) \rightarrow (HCOO^-)_2Ca^{2+}(aq) + CO_2(g) + H_2O(l)$ *[1]*

c $2CH_3CH_2CH_2COOH(aq) + Zn(s) \rightarrow (CH_3CH_2CH_2COO^-)_2Zn^{2+}(aq) + H_2(g)$ *[1]*

3 a $HOOC–COOH(aq) + Na_2CO_3(aq) \rightarrow Na^+(^-OOC–COO^-)Na^+(aq) + CO_2(g) + H_2O(l)$ *[1]*

b $HOOC–COOH(aq) + NaOH(aq) \rightarrow HOOC–COO^-Na^+(aq) + H_2O(l)$ *[1]*

26.4

1 a methyl butanoate *[1]*

b pentyl propanoate *[1]*

c propyl methanoate *[1]*

2 a i $CH_3CH_2CH_2COOH + CH_3CH_2CH_2OH \rightarrow CH_3CH_2CH_2COOCH_2CH_2CH_3 + H_2O$ *[1]*

ii $HCOOH + CH_3(CH_2)_4CH_2OH \rightarrow HCOOCH_2(CH_2)_4CH_3 + H_2O$ *[1]*

b $CH_3CH_2COOH + SOCl_2 \rightarrow CH_3CH_2COCl + SO_2 + HCl$ *[1]*

c i $CH_3CH_2COOCH_2CH_2CH_3 + H_2O \rightarrow CH_3CH_2COOH + CH_3CH_2CH_2OH$ *[1]*

ii $CH_3CH_2COOCH_2CH_2CH_3 + OH^- \rightarrow CH_3CH_2COO^- + CH_3CH_2CH_2OH$ *[1]*

3 a $C_6H_5COCl + CH_3CH_2OH \rightarrow C_6H_5COOCH_2CH_3 + HCl$ *[1]*

b $(CH_3CH_2CO)_2O + CH_3(CH_2)_4OH \rightarrow CH_3CH_2COO(CH_2)_4CH_3 + CH_3CH_2COOH$ *[1]*

c $CH_3CH_2CH_2COOH + SOCl_2 \rightarrow CH_3CH_2CH_2COCl + SO_2 + HCl$ *[1]*

$CH_3CH_2CH_2COCl + 2NH_3 \rightarrow CH_3CH_2CH_2CONH_2 + NH_4Cl$ *[1]*

27.1

1 a The lone pair of electrons on the N atom can accept a proton. *[1]*

A dative covalent bond forms between the lone pair of electrons on the N atom and H^+. *[1]*

b i $C_6H_5NH_2 + HCl \rightarrow C_6H_5NH_3^+Cl^-$ *[1]*

ii phenylammonium chloride *[1]*

2 a chlorobutane *[1]*; Step 1: excess ammonia in ethanol *[1]*, Step 2: NaOH(aq) *[1]*

b $C_4H_9NH_3^+Cl^-$ *[1]*

3 a $2C_6H_5NH_2 + H_2SO_4 \rightarrow (C_6H_5NH_3^+)_2SO_4^{2-}$ *[1]*

b *[2]*

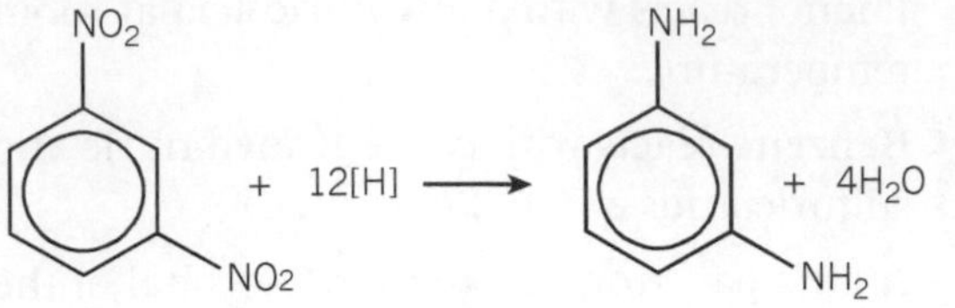

27.2

1 a i 1 *[1]*; **ii** 2 *[1]*

b There are only 3 different groups attached to the α-carbon. *[1]*

2 a i $HSCH_2CH(NH_2)COOH$ *[1]*

ii $(CH_3)_2CHCH(NH_2)COOH$ *[1]*

b **i**

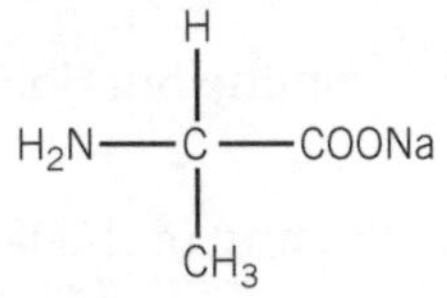

ii

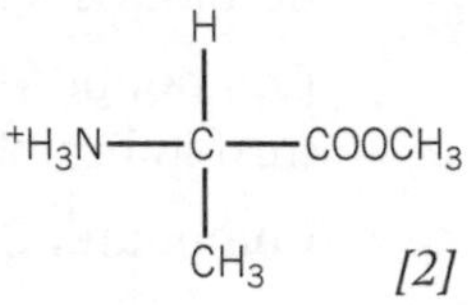

[2]

3 a i

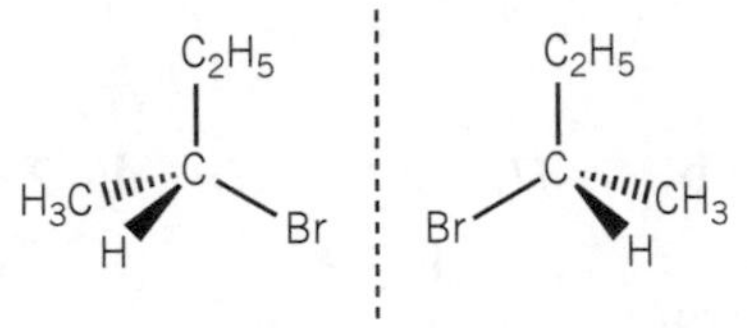

ii

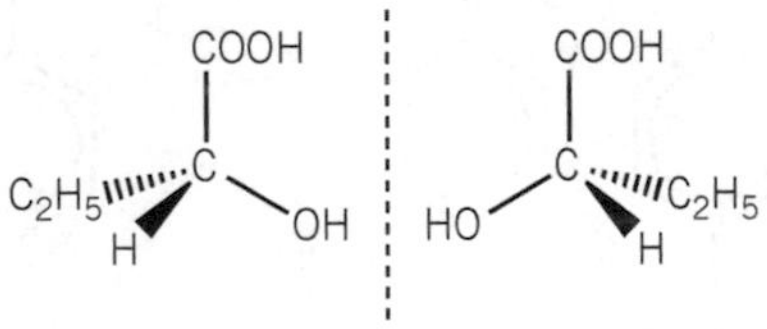

[4]

b **i**

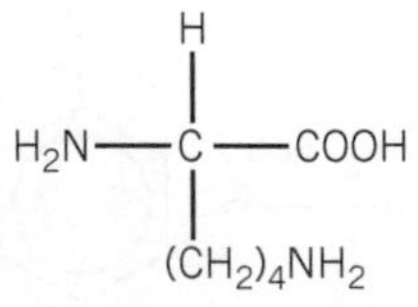

ii

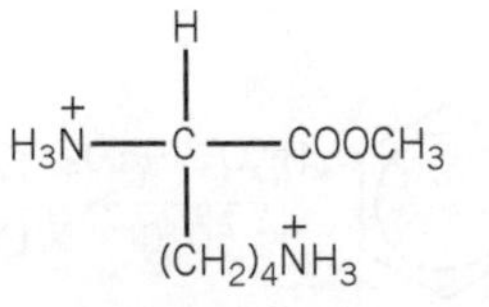

[3]

27.3

1 a i Carboxylic acid and amine. *[1]*

ii Carboxylic acid and alcohol/hydroxyl. *[1]*

b i Formation of a very long molecular chain, by repeated addition reactions of many unsaturated alkene molecules (monomers). *[1]*

ii A reaction in which small molecules react together to form a very long molecular chain with elimination of small molecules such as water. *[1]*

2 a i

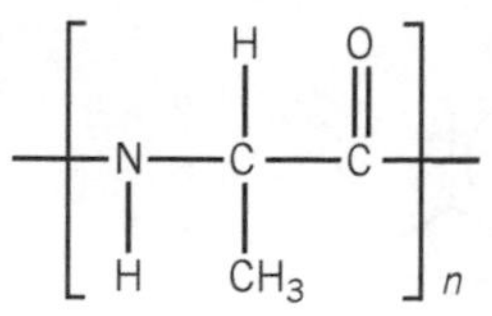

[2]

ii

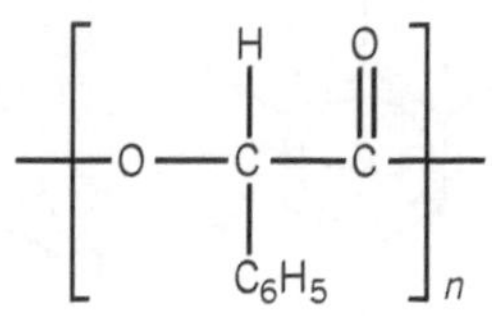

[2]

iii

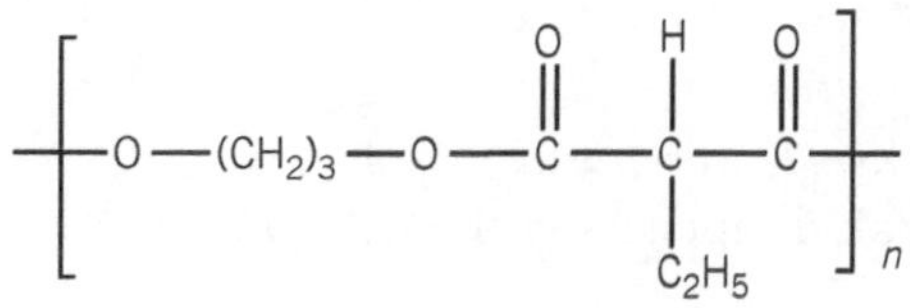

[2]

iv

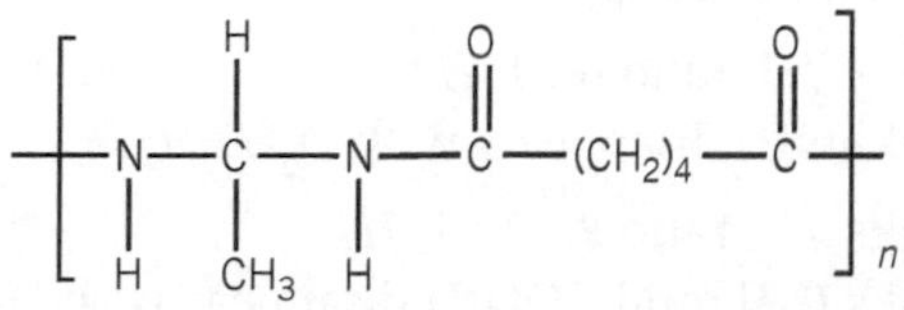

[2]

3 a

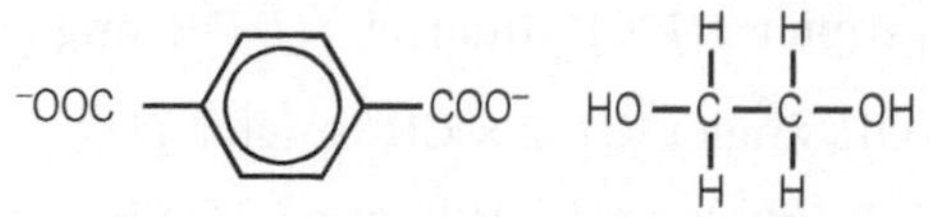

[2]

b

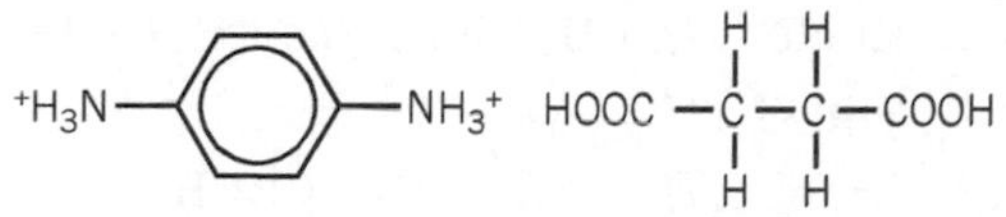

[2]

28.1

1 $CH_3CH_2Cl + KCN \rightarrow CH_3CH_2CN + KCl$ *[1]*

$CH_3CHClCH_3 + KCN \rightarrow CH_3CH(CN)CH_3 + KCl$ *[1]*

1-chloropropane increases carbon chain length. 2-chloropropane adds a carbon side chain. *[1]*

2 $CH_3CH_2CHO + HCN \rightarrow CH_3CH_2CH(OH)CN$ *[1]*

Reagents and conditions: $NaCN/H_2SO_4$ *[1]*

$CH_3CH_2CH(OH)CN + 2H_2O + HCl \rightarrow CH_3CH_2CH(OH)COOH + NH_4Cl$ *[1]*

Reagents and conditions: HCl(aq) and heat *[1]*

$CH_3CH_2CH(OH)COOH + NaOH \rightarrow CH_3CH_2CH(OH)COONa + H_2O$ *[1]*

Reagents and conditions: NaOH(aq) *[1]*

3 a $C_6H_6 + (CH_3)_2CHCl \rightarrow C_6H_5CH(CH_3)_2 + HCl$ *[1]*

Reagents and conditions: $(CH_3)_2CHCl$ and $AlCl_3$ halogen carrier catalyst. *[1]*

b $C_6H_6 + HCOCl \rightarrow C_6H_5CHO + HCl$ *[1]*

Reagents and conditions: HCOCl and $AlCl_3$ halogen carrier catalyst. *[1]*

$C_6H_5CHO + [O] \rightarrow C_6H_5COOH$ *[1]*

Reagents and conditions: $K_2Cr_2O_7/H_2SO_4$, heat. *[1]*

c $C_6H_6 + CH_3CH_2COCl \rightarrow C_6H_5COCH_2CH_3 + HCl$ *[1]*

Reagents and conditions: CH_3CH_2COCl and $AlCl_3$ halogen carrier catalyst. *[1]*

$C_6H_5COCH_2CH_3 + 2[H] \rightarrow C_6H_5CH(OH)CH_2CH_3$ *[1]*

Reagents and conditions: $NaBH_4$. *[1]*

28.2

1 Buchner flask and funnel, filter paper and access to a vacuum pump or filter pump connected to a water tap. *[2]*

2 A pure organic substance usually has a very sharp melting point. Impure organic compounds have lower melting points, and melt over a wider temperature range, than the pure compound. *[2]*

3 Filtration under reduced pressure removes the impure organic solid from the reaction mixture. *[1]*

Recrystallisation removes soluble impurities from the desired product. *[1]*

Measurement of the melting point checks the purity of the recrystallised sample. *[1]*

28.3

Figure 2: From top,

ketone, alkene, and secondary alcohol (nandrolone, a steroid)

phenol, secondary alcohol, secondary amine (adrenaline, a hormone)

arene, ester, secondary amide, primary amine, carboxylic acid (Aspartame, a sweetener)

1 **A**: Alkene and ester **B**: Acyl chloride and phenol *[2]*

2 Step 1 – HBr *[1]*

$H_2C{=}CH_2 + HBr \rightarrow CH_3CH_2Br$ *[1]*

Step 2 – Excess NH_3/ethanol *[1]*

$CH_3CH_2Br + 2NH_3 \rightarrow CH_3CH_2NH_2 + NH_4Cl$ *[1]*

3 Step 1 – Aqueous sodium hydroxide *[1]*

$CH_3CH_2Br + NaOH \rightarrow CH_3CH_2OH + NaBr$ *[1]*

Divide C_2H_5OH sample in two.

Step 2 – Acidified potassium dichromate / reflux *[1]*

$CH_3CH_2OH + 2[O] \rightarrow CH_3COOH + H_2O$ *[1]*

Step 3 – React CH_3COOH and C_2H_5OH with acid catalyst *[1]*

$CH_3COOH + C_2H_5OH \rightarrow CH_3COOC_2H_5 + H_2O$ *[1]*

29.1

1 a TLC: The solvent *[1]* GC : The gas *[1]*

b i Components are separated by their relative adsorptions to the solid stationary phase. *[1]*

ii Components are separated by their relative solubility in the liquid stationary phase. *[1]*

2 a A: $R_f = 0.53$ *[1]*, B: $R_f = 0.37$ *[1]*, C: $R_f = 0.24$ *[1]*.
A is isoleucine, B is cysteine, C is aspartic acid *[1]*

b i Add $Na_2CO_3(aq)$. The carboxylic acid effervesces, the phenol does not. *[1]*

ii $2RCOOH + Na_2CO_3 \rightarrow 2RCOONa + CO_2 + H_2O$ *[1]*

3 From $H^+/Cr_2O_7^{2-} \rightarrow$ green, A, B, and C are primary alcohol, secondary alcohol, and aldehyde. *[1]*

From 2,4-DNP result, A is an aldehyde or ketone. *[1]*

From $Na_2CO_3(aq)$ result, A and B react with $H^+/Cr_2O_7^{2-}$ to form a carboxylic acid. *[1]*

Therefore A is an aldehyde. *[1]*

Therefore B must be a primary alcohol and C is a secondary alcohol (not oxidised to –COOH). *[1]*

29.2

1 ^1H, ^{13}C, ^{15}N, ^{31}P *[1]*

2 a An example is $CDCl_3$. *[1]*

Deuterated solvents do not produce peaks in ^1H NMR spectroscopy. *[1]*

b i DMSO contains ^1H atoms which will produce a peak in a ^1H NMR spectrum. *[1]*

ii Replace the ^1H atoms with deuterium and use as $(CD_3)_2SO$. *[1]*

3 a Chemical shift is the shift in frequency, compared to TMS, required for nuclear magnetic resonance to take place. *[1]*

The shift depends on the chemical environment, especially caused by electronegative atoms or π-bonds. *[1]*

b i $(CH_3)_4Si$ *[1]*

ii The protons are all in the same chemical environment. *[1]*

29.3

1 a $CH_3CH(OH)CH_3$: 2 peaks, 2 × CH_3–C at 0–50 ppm; CHOH at 50–90 ppm *[2]*

b CH_3COOCH_3: 3 peaks, CH_3–C at 0–50 ppm; OCH_3 at 50–90 ppm; CO at 160–220 ppm *[2]*

c $CH_3CH_2COCH_2CH_3$: 3 peaks, 2 × CH_3–C at 0–50 ppm; 2 × CH_2–C at 0–50 ppm; CO at 160–220 ppm *[2]*

d C_6H_5OH: 4 aromatic peaks in the range 110–160 ppm *[2]*

2 $CH_3CH_2CH_2CH_2Cl$: 4 peaks *[1]*, $(CH_3)_2CHCH_2Cl$: 3 peaks *[1]*
$CH_3CH_2CH(CH_3)Cl$: 4 peaks *[1]*, $(CH_3)_3CCl$: 2 peaks *[1]*

A is $(CH_3)_2CHCH_2Cl$ and **B** is $(CH_3)_3CCl$ *[1]*

3

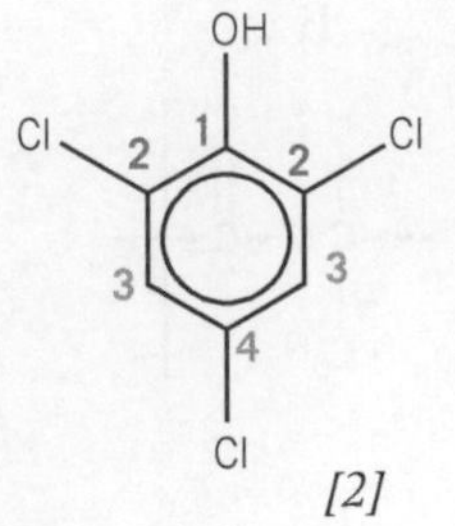

[2]

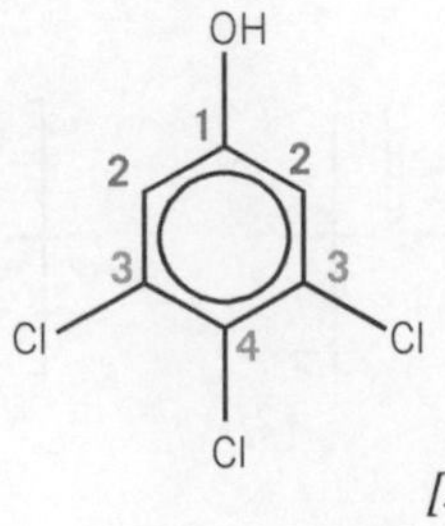

[2]

29.4

1 a 4 peaks *[1]*, ratio 3 : 2 : 2 : 1 *[1]*, no peak disappears with D_2O. *[1]*

b 4 peaks *[1]*, ratio 6 : 1 : 2 : 1 *[1]*, 1 peak (OH) disappears with D_2O. *[1]*

c 2 peaks *[1]*, ratio 6 : 1 *[1]*, 1 peak (NH) disappears with D_2O. *[1]*

d 3 peaks *[1]*, ratio 2 : 2 : 1 *[1]*, 2 peaks (NH_2 and COOH) disappear with D_2O. *[1]*

2 a CH_3 triplet *[1]*, CH_2 quartet *[1]*, OH singlet *[1]*

b 2 × OH singlet *[1]*, 2 × CH_2 singlet *[1]*

c $(CH_3)_2$ doublet *[1]*, CH heptet *[1]*, NH_2 singlet *[1]*

d OH, singlet *[1]*, CH_2 triplet *[1]*, CH_2 triplet *[1]*, CH_3 singlet *[1]* *[1]*

3 a CH_3CH_2 *[1]* b $CHCH_2$ *[1]*

c CHCH *[1]* d $(CH_3)_2CH$ *[1]*

29.5

1 2 peaks at δ = 0–50 ppm for 2 different **C**–C *[1]*
1 peak at δ = 50–90 ppm for **C**–O *[1]*
$(CH_3)_3CCH_2OH$ *[1]*

2 a Triplet (3H) at 1.2 ppm, $\mathbf{CH_3}CH_2$ *[1]*
Quartet (2H) at 2.2 ppm, $CH_3\mathbf{CH_2}CO$ *[1]*
Singlet (3H) at 3.8 ppm, $O\mathbf{CH_3}$ *[1]*
$CH_3CH_2COOCH_3$ *[1]*

b Triplet (2H) at 2.2 ppm, $CH_2\mathbf{CH_2}CO$ *[1]*
Triplet (2H) at 3.8 ppm, $O\mathbf{CH_2}CH_2$ *[1]*
1H singlet at 4.5 ppm, O**H** *[1]*
1H singlet at 11.5 ppm, COO**H** *[1]*
$HOCH_2CH_2COOH$ *[1]*

c Singlet (9H) at 1.2 ppm, $(\mathbf{CH_3})_3C$ *[1]*
Doublet (2H) at 2.2 ppm, $C\mathbf{CH_2}CHO$ *[1]*
Triplet (1H) at 9.5 ppm, $CH_2C\mathbf{H}O$ *[1]*
$(CH_3)_3CCH_2CHO$ *[1]*

3 Heptet at δ = 2.7 ppm, CH adjacent to $(CH_3)_2$ and C=O *[1]*
Doublet at δ =1.1 ppm, $(CH_3)_2$ adjacent to CH *[1]*
Singlet at δ =2.0 ppm, CH_3 adjacent to C=O *[1]*
$(CH_3)_2CHCOCH_3$ *[1]*

29.6

1 C : H : O = 58.83/12.0 : 9.80/1.0 : 31.37/16.0 = 4.90 : 9.80 : 1.96 = 5 : 10 : 2 *[1]*

Empirical formula = $C_5H_{10}O_2$ *[1]*
(M = 60 + 10 + 32 = 102)

Molecular formula = $C_5H_{10}O_2$ × 102/102 = $C_5H_{10}O_2$ *[1]*

2 Peak at 1740 cm^{-1} shows C=O group in an aldehyde, ketone, carboxylic acid, or ester. *[1]*

No broad peak at 2500–3300 cm^{-1}, so not a COOH; also no alcohol OH peak. *[1]*

3 a 3 peaks → 3 types of proton *[1]*

peak area ratio → 1 : 3 : 6 (from left)
→ CH : CH_3 : $(CH_3)_2$ *[1]*

CH heptet at 4.9 ppm
→ CH adjacent to $(CH_3)_2$ and –O: O–CH$(CH_3)_2$

CH_3 singlet at 2.2 ppm
→ CH_3 adjacent to C=O: CH_3CO

$(CH_3)_2$ doublet at 1.3 ppm
→ $(CH_3)_2$ adjacent to CH: CH$(CH_3)_2$ *[3]*

Unknown compound: $CH_3COOCH(CH_3)_2$ *[1]*

b There are 4 carbon environments and 4 peaks. *[1]*

2 peaks at 0–50 ppm for **C**H_3–C and C–(**C**H_3)$_2$ environments *[1]*

1 peak at 50–90 ppm for –**C**–O environment *[1]*

1 peak at 160–220 ppm for **C**=O environment *[1]*

Answers to practice questions

Chapter 18

1 D *[1]*; **2** C *[1]*; **3** A *[1]*; **4** C *[1]*; **5** A *[1]*.

6 a Order with respect to **A** = 1st order *[1]*

From Experiment 2 to 3, [**A**] increases by 1.5 and rate increases by 1.5 *[1]*

Order with respect to **B** = 2nd order *[1]*

From Experiment 1 to 2, [**A**] and [**B**] both double and rate increases by 8

[**A**] is 1st order so effect of [**B**] is to increase rate by 4 times = 2^2 *[1]*

b rate = $k[\mathbf{A}][\mathbf{B}]^2$ *[1]*

c $k = \frac{\text{rate}}{[\mathbf{A}][\mathbf{B}]^2} = \frac{1.5 \times 10^{-4}}{0.0100 \times 0.0100^2}$

$= 150\ \text{dm}^6\,\text{mol}^{-2}\,\text{s}^{-1}$ *[2]*

d A

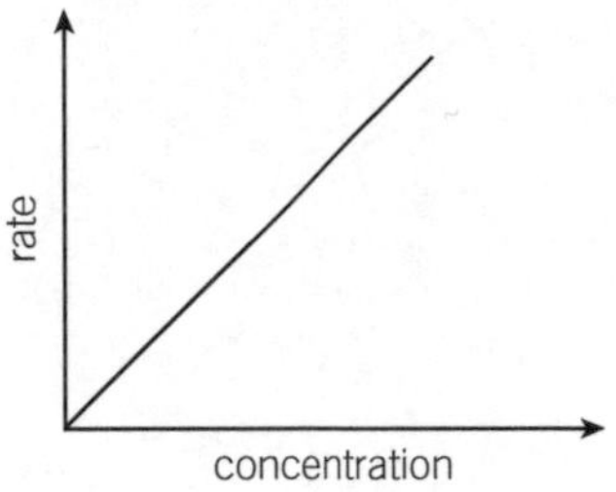

[1]

B

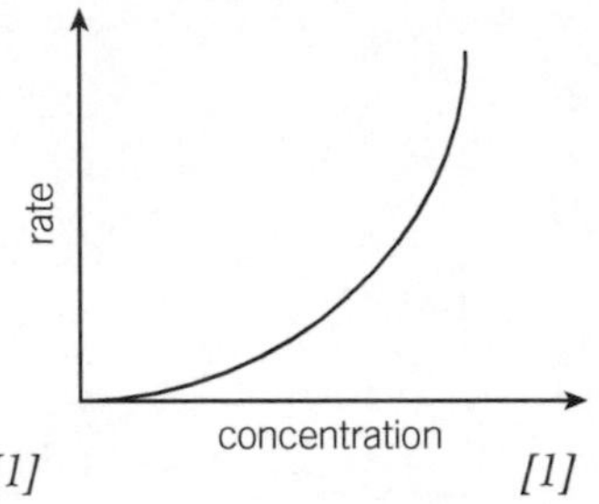

[1]

7 a rate = $k[NO][O_3]$ *[1]*

b $O_3 + O \rightarrow 2O_2$ *[1]*

c NO is a catalyst. *[1]* NO reacts in Step 1 and is regenerated in Step 2. *[1]*

Chapter 19

1 C *[1]*; **2** C *[1]*; **3** D *[1]*; **4** D *[1]*

5 a $[CO] = (0.100 - 0.080) \times 2 = 0.040\ \text{mol dm}^{-3}$ *[1]*

$[H_2] = (0.200 - 0.160) \times 2 = 0.08\ \text{mol dm}^{-3}$ *[1]*

$[CH_3OH] = 0.080 \times 2 = 0.160\ \text{mol dm}^{-3}$ *[1]*

b $K_c = \frac{[CH_3OH(g)]}{[CO(g)][H_2(g)]^2}$ *[1]*

c $K_c = \frac{0.160}{0.040 \times 0.080^2}$ [1] $= 625\ \text{dm}^6\,\text{mol}^{-2}$ *[1]*

d The system is no longer in equilibrium as $\frac{\text{products}}{\text{reactants}}$ is now **less** than K_c. *[1]*

Products increase and reactants decrease until $\frac{\text{products}}{\text{reactants}}$ is once again equal to K_c *[1]*

Equilibrium position shifts to the right. *[1]*

6 a $p(SO_2) = \left(\frac{0.15}{1.5}\right) \times 200 = 20\ \text{kPa}$ *[1]*

$p(O_2) = \left(\frac{0.45}{1.5}\right) \times 200 = 60\ \text{kPa}$ *[1]*

$p(SO_3) = \left(\frac{0.90}{1.5}\right) \times 200 = 120\ \text{kPa}$ *[1]*

b $K_p = \frac{p(SO_3)^2}{p(SO_2)^2 \times p(O_2)}$ *[1]*

c $K_p = \frac{120^2}{20^2 \times 60} = 0.600\ \text{kPa}^{-1}$ *[2]*

d The forward reaction is exothermic and K_p decreases. *[1]*

The system is no longer in equilibrium as $\frac{\text{products}}{\text{reactants}}$ is now **greater** than K_p. *[1]*

Equilibrium position shifts to the left, products decrease and reactants increase until $\frac{\text{products}}{\text{reactants}}$ is equal to the new value of K_p *[1]*

Chapter 20

1 B *[1]*; **2** A *[1]*; **3** B *[1]*; **4** D *[1]*

5 a i $pH = -\log(0.125) = 0.090$ *[1]*

ii $[H^+(aq)] = \frac{K_w}{[OH^-(aq)]} = \frac{1.00 \times 10^{-14}}{0.125}$

$= 8.00 \times 10^{-14}\ \text{mol dm}^{-3}$

$pH = -\log(8.00 \times 10^{-14}) = 13.10$ *[2]*

b [HCl] in diluted solution = $0.0625\ \text{mol dm}^{-3}$ *[1]*

$pH = -\log(0.0625) = 1.20$ *[1]*

c $n(NaOH) = 0.0800 \times \frac{20.0}{1000} = 1.60 \times 10^{-3}\ \text{mol}$ *[1]*

$n(HCl) = 0.0600 \times \frac{80.0}{1000} = 4.80 \times 10^{-3}\ \text{mol}$ *[1]*

$n(HCl)$ remaining $= 4.80 \times 10^{-3} - 1.60 \times 10^{-3}$

$= 3.20 \times 10^{-3}$ mol in 100 cm^3 *[1]*

$[HCl] = 3.20 \times 10^{-2}\ \text{mol dm}^{-3}$ *[1]*

$pH = -\log 3.20 \times 10^{-2} = 1.49$ *[1]*

6 a i A proton donor *[1]*

ii Partially dissociates releasing H^+ *[1]*

b $CH_3CH_2COOH(aq) + H_2O(l) \rightleftharpoons$
Acid 1 Base 2

$H_3O^+(aq) + CH_3CH_2COO^-(aq)$ *[1]*
Acid 2 Base 1 *[1]*

c $2CH_3CH_2COOH(aq) + CaCO_3(s) \rightarrow (CH_3CH_2COO)_2Ca(aq) + CO_2(g) + H_2O(l)$ *[1]*

$2H^+(aq) + CaCO_3(s) \rightarrow Ca^{2+}(aq) + CO_2(g) + H_2O(l)$ *[1]*

d i $K_a = \frac{[H^+(aq)]\,[CH_3CH_2COO^-(aq)]}{[CH_3CH_2COOH(aq)]}$ *[1]*

ii $K_a = 10^{-4.87} = 1.35 \times 10^{-5}\ \text{mol dm}^{-3}$ *[1]*

$[H^+(aq)] = \sqrt{(1.35 \times 10^{-5} \times 0.125)}$

$= 1.30 \times 10^{-3}\ \text{mol dm}^{-3}$ *[1]*

$pH = -\log 1.30 \times 10^{-3} = 2.89$ *[1]*

iii The approximation $H^+(aq) \sim [A^-(aq)]$ breaks down for very weak acids and very dilute solutions.

The approximation
$[HA]_{equilibrium} \sim [HA]_{undissociated}$ breaks down for 'stronger' weak acids and concentrated solutions. *[2]*

Chapter 21

1 D *[1]*; **2** B *[1]*; **3** C *[1]*; **4** B *[1]*.

5 a Strong acid–weak base starts at pH ~ 1 and finishes at pH ~ 12; *[1]*

weak acid–strong base starts at pH ~ 3 and finishes at pH ~ 14. *[1]*

b i Indicator needs to change colour within the pH range of the vertical section of the curve. *[1]*

The curve moves from one end of the vertical section to the other end with addition of a few drops of base. *[1]*

ii Strong acid–weak base: 2,4-nitrophenol and cresol red; *[1]*

weak acid–strong base: cresol red and thymolphthalein. *[1]*

6 a i Carbonic acid **ii** hydrogencarbonate ion *[2]*

b $K_a = \dfrac{[H^+(aq)]\,[HCO_3^-(aq)]}{[H_2CO_3(aq)]}$ *[1]*

c On addition of an acid, $H^+(aq)$ ions react with $HCO_3^-(aq)$. *[1]*

The equilibrium position shifts to the left, removing most of the $H^+(aq)$ ions. *[1]*

On addition of an alkali, $OH^-(aq)$, the small concentration of $H^+(aq)$ ions reacts with the $OH^-(aq)$ ions:

$H^+(aq) + OH^-(aq) \rightarrow H_2O(l)$. *[1]*

$H_2CO_3(aq)$ dissociates and the equilibrium position shifts to the right, restoring most of the $H^+(aq)$ ions. *[1]*

7 a $[H^+(aq)] = 1.70 \times 10^{-4} \times \dfrac{0.250}{0.450}$

$= 9.44 \times 10^{-5}\,mol\,dm^{-3}$ *[1]*

$pH = -\log(9.44 \times 10^{-5}) = 4.02$ *[1]*

b $[HCOOH] = 0.250 \times \dfrac{25.0}{100} = 0.0625\,mol\,dm^{-3}$ *[1]*

$[HCOO^-] = 0.200 \times \dfrac{75}{100} = 0.150\,mol\,dm^{-3}$ *[1]*

$[H^+(aq)] = 1.70 \times 10^{-4} \times \dfrac{0.0625}{0.150}$

$= 7.083 \times 10^{-5}\,mol\,dm^{-3}$ *[1]*

$pH = -\log(7.083 \times 10^{-5}) = 4.15$ *[1]*

c $n(HCOO^-) = 0.250 \times \dfrac{400}{1000} = 0.100\,mol$ *[1]*

$n(HCOOH) = 0.400 \times \dfrac{600}{1000} - 0.100$

$= 0.240 - 0.100 = 0.140\,mol$ *[1]*

Total volume = $1\,dm^3$ and so concentrations in $mol\,dm^{-3}$ = amount in moles

$[H^+(aq)] = 1.70 \times 10^{-4} \times \dfrac{0.140}{0.100}$

$= 2.38 \times 10^{-4}\,mol\,dm^{-3}$ *[1]*

$pH = -\log(2.38 \times 10^{-4}) = 3.62$ *[1]*

Chapter 22

1 D *[1]*; **2** A *[1]*; **3** D *[1]*; **4** B *[1]*.

5 a i Lattice enthalpy is the enthalpy change that accompanies the formation of one mole of an ionic compound from its gaseous ions under standard conditions. *[1]*

ii $Ca^{2+}(g) + 2F^-(g) \rightarrow CaF_2(s)$ *[1]*

b i

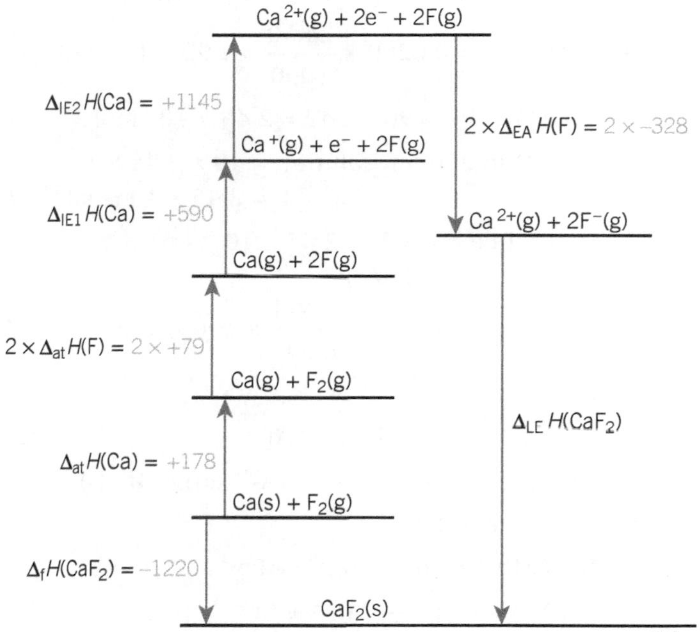

[5]

ii $178 + 2 \times 79 + 590 + 1145 + 2 \times (-328) + \Delta_{LE}H^{\ominus}(CaF_2) = -1220$ *[1]*

$\therefore \Delta_{LE}H^{\ominus}(CaF_2) = -1220 - 1415$

$= -2635\,kJ\,mol^{-1}$ *[1]*

c Lattice enthalpy depends upon ionic size and ionic charge. *[1]*

As ionic size increases, attraction between oppositely charged ions decreases and lattice enthalpy becomes less exothermic. *[1]*

As ionic charge increases, attraction between oppositely charged ions increases and lattice enthalpy becomes more exothermic. *[1]*

6 a $\Delta S = [\,(2 \times 27.3) + (3 \times 197.6)\,] - [\,87.4 + (3 \times 5.7)\,] = +542.9\,J\,mol^{-1}\,K^{-1}$ *[1]*

$\Delta H = [\,(3 \times -110.5)\,] - (-824.2)$

$= +492.7\,kJ\,mol^{-1}$ *[1]*

b $\Delta S = 542.9/1000 = 0.5429\,kJ\,mol^{-1}\,K^{-1}$

$\Delta G = 492.7 - 298 \times 0.5429 = +330.9\,kJ\,mol^{-1}$ *[1]*

As $\Delta G > 0$, the reaction is not feasible at 25 °C *[1]*

c For minimum temperature $\Delta G = 0$; $\Delta H - T\Delta S = 0$:

$T = \frac{\Delta H}{\Delta S}$ *[1]*

Minimum temperature $T = \frac{492.7}{0.5429} = 907.5\,\text{K}$

$= 634.5\,^\circ\text{C}$ *[1]*

Chapter 23

1 A *[1]*; **2** B *[1]*; **3** A *[1]*

4 a i $Cu + 2NO_3^- + 4H^+ \rightarrow Cu^{2+} + 2NO_2 + 2H_2O$ *[1]*

ii $2MnO_4^- + 16H^+ + 5Sn^{2+} \rightarrow 2Mn^{2+} + 5Sn^{4+} + 8H_2O$ *[1]*

b $Cr_2O_7^{2-} + 6I^- + 14H^+ \rightarrow 2Cr^{3+} + 3I_2 + 7H_2O$ *[2]*

5 a $5Fe^{2+}(aq) + MnO_4^-(aq) + 8H^+(aq) \rightarrow 5Fe^{3+}(aq) + Mn^{2+}(aq) + 4H_2O(l)$ *[1]*

b $n(MnO_4^-) = 0.0200 \times \frac{24.50}{1000} = 4.90 \times 10^{-4}\,\text{mol}$ *[1]*

$n(Fe^{2+}) = \mathbf{5} \times 4.90 \times 10^{-4} = 2.45 \times 10^{-3}\,\text{mol}$ *[1]*

$n(Fe^{2+})$ in $250\,\text{cm}^3$ solution $= \mathbf{10} \times 2.45 \times 10^{-3} = 2.45 \times 10^{-2}\,\text{mol}$ *[1]*

mass of Fe $= n \times M = 2.45 \times 10^{-2} \times 55.8 = 1.3671\,\text{g}$ *[1]*

Percentage purity $= \frac{1.3671}{6.84} \times 100 = 20.0\%$ *[1]*

6 a i Mn^{3+} **ii** Al *[2]*

b i 1.22 V **ii** $Al^{3+}|Al$ *[2]*

iii $2Al(s) + 3Fe^{2+}(aq) \rightarrow 2Al^{3+}(aq) + 3Fe(s)$ *[1]*

c i Cr^{2+} and Pb *[1]*

ii $2Mn^{3+} + Pb \rightarrow 2Mn^{2+} + Pb^{2+}$ *[1]*

$Mn^{3+} + Cr^{2+} \rightarrow Mn^{2+} + Cr^{3+}$ *[1]*

Chapter 24

1 B *[1]*; **2** C *[1]*; **3** A *[1]*

4 a i $[Fe(CN)_4(NH_3)_2]^-$; $FeC_4N_6H_6^-$ *[2]*

ii

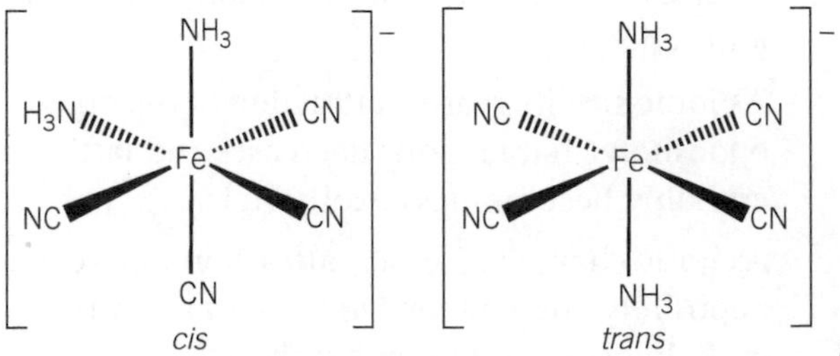

cis trans

[2]

b i $[Cr(H_2NCH_2CH_2NH_2)_2F_2]^+$; $CrN_4C_4H_{16}F_2^+$ *[2]*

ii

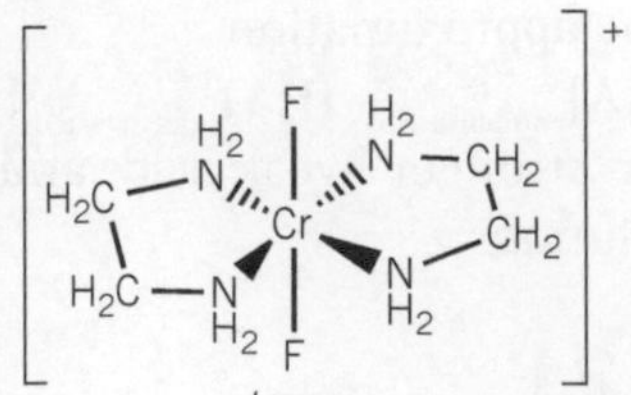

trans

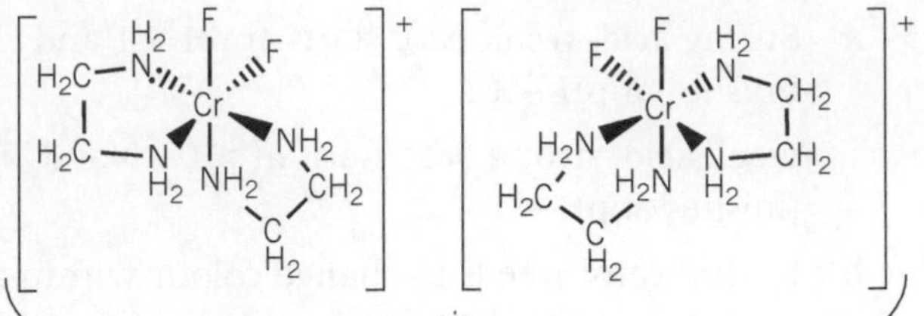

cis

optical *[3]*

5 a **M** = Fe; **A** = $[Fe(H_2O)_6]^{2+}$ *[2]*

b $Fe^{2+}(aq) + 2OH^-(aq) \rightarrow Fe(OH)_2(s)$ *[1]*

c **B** = $Fe(OH)_3$

Formed from oxidation of $Fe(OH)_2$ by air. *[2]*

6 a **A:** violet, $[Cr(H_2O)_6]^{3+}$ *[2]* **B:** green, $Cr(OH)_3$ *[2]*

C: dark green $[Cr(OH)_6]^{3-}$ *[2]*

b i $Cr^{3+}(aq) + 3OH^-(aq) \rightarrow Cr(OH)_3(s)$ *[1]*

ii $Cr(OH)_3(s) + 3OH^-(aq) \rightarrow [Cr(OH)_6]^{3-}(aq)$ *[1]*

iii $[Cr(H_2O)_6]^{3+} + 6OH^-(aq) \rightarrow [Cr(OH)_6]^{3-} + 6H_2O$ *[1]*

Chapter 25

1 D *[1]*; **2** D *[1]*; **3** C *[1]*; **4** C *[1]*

5 a i 1,2-dibromocyclohexane *[1]*

electrophilic addition *[1]*

ii 2,4,6-tribromophenol *[1]*

electrophilic substitution. *[1]*

b i $AlBr_3/FeBr_3/Fe$ *[1]*

ii $Br_2 + AlBr_3 \rightarrow Br^+ + AlBr_4^-$ *[1]*

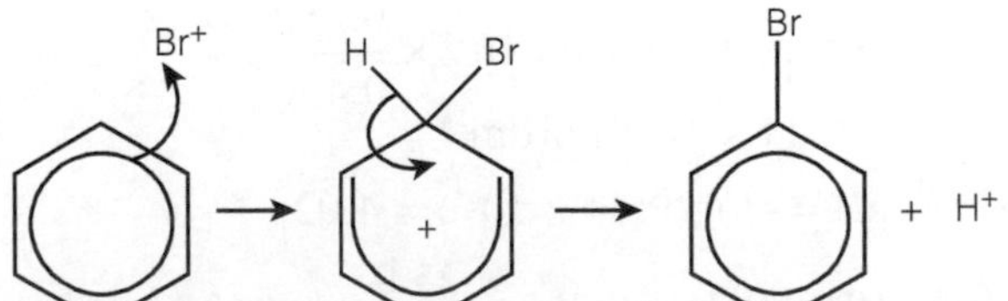

[3]

$AlBr_4^- + H^+ \rightarrow AlBr_3 + HBr$ *[1]*

c A lone pair from the oxygen p-orbital of the –OH group is donated into the π-system of phenol. *[1]*

The electron density of the aromatic ring in phenol increases. *[1]*

The increased electron density attracts electrophiles more strongly than with benzene. *[1]*

6 a $C_6H_5OH \rightleftharpoons C_6H_5O^- + H^+$ *[1]*

b i $C_6H_5OH + NaOH \rightarrow C_6H_5O^-Na^+ + H_2O$ *[1]*

ii sodium phenoxide; neutralisation. *[2]*

c $2C_6H_5OH + Mg \rightarrow (C_6H_5O)_2Mg + H_2$ *[2]*
redox reaction. *[1]*

d i 2-nitrophenol; 4-nitrophenol *[2]*

ii

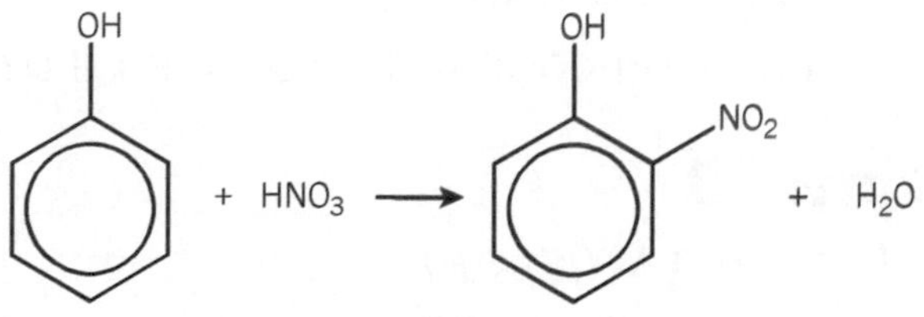

(or 4-nitro product) *[2]*

iii 2,4,6-trinitrophenol *[1]*

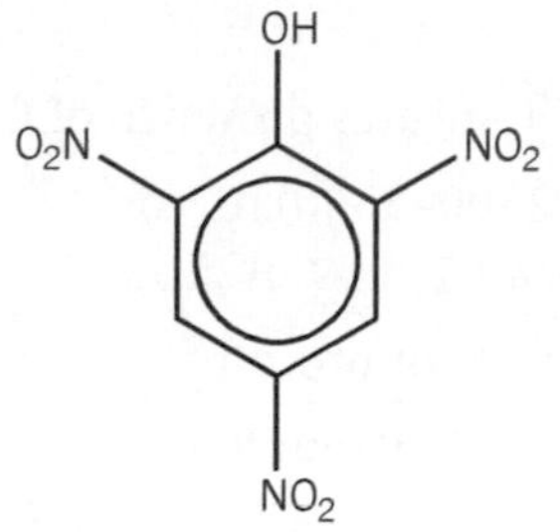

[1]

Chapter 26

1 A *[1]*; **2** A *[1]*; **3** C *[1]*; **4** B *[1]*

5 a i $H_2SO_4/K_2Cr_2O_7$ *[1]*
$CH_3CH_2COCHO + [O] \rightarrow CH_3CH_2COCOOH$ *[2]*

ii $NaBH_4$ *[1]*
$CH_3CH_2COCHO + 4[H] \rightarrow$
$CH_3CH_2CH(OH)CH_2OH$ *[2]*

b i *[1]*

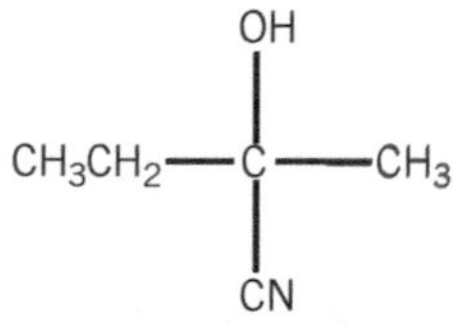

ii hydroxyl/ alcohol and nitrile *[2]*

iii nucleophilic addition *[1]*

iv

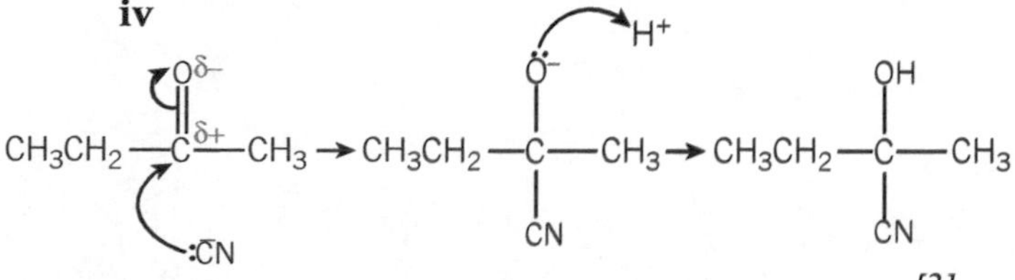

[3]

c Acid: $CH_3COCH_2COOCH_3 + H_2O \rightarrow$
$CH_3COCH_2COOH + CH_3OH$ *[2]*
Alkali: $CH_3COCH_2COOCH_3 + OH^- \rightarrow$
$CH_3COCH_2COO^- + CH_3OH$ *[2]*

6 a i $CH_3CH_2COOH + CH_3(CH_2)_4OH \rightarrow$
$CH_3CH_2COO(CH_2)_4CH_3 + H_2O$ *[2]*

ii $CH_3COOH + CH_3CH_2CH(CH_3)OH \rightarrow$
$CH_3COOCH(CH_3)CH_2CH_3 + H_2O$ *[2]*

iii $C_6H_5COOH + C_6H_5CH_2OH \rightarrow$
$C_6H_5COOCH_2C_6H_5 + H_2O$ *[2]*

b i $2CH_3CH_2COOH + CaCO_3 \rightarrow$
$(CH_3CH_2COO)_2Ca + CO_2 + H_2O$ *[1]*

ii $(CH_3CO)_2O + (CH_3)_2CHOH \rightarrow$
$CH_3COOCH(CH_3)_2 + CH_3COOH$ *[2]*

iii $CH_3COCl + 2NH_3 \rightarrow CH_3CONH_2 + NH_4^+Cl^-$ *[2]*

c Concentrated H_2SO_4 catalyst *[1]*
$HOCH_2CH_2CH_2COOH$ *[1]*

Chapter 27

1 D *[1]*; **2** D *[1]*; **3** A *[1]*

4 a i The lone pair of electrons on the N atom can accept a proton. *[1]*
A dative covalent bond forms between the lone pair of electrons on the N atom and H^+. *[1]*

ii $2C_6H_5NH_2 + H_2SO_4 \rightarrow (C_6H_5NH_3^+)_2SO_4^{2-}$ *[2]*

b i 1-Bromopropane and excess ammonia in ethanol *[1]*

ii $CH_3CH_2CH_2Br + 2NH_3 \rightarrow$
$CH_3CH_2CH_2NH_2 + NH_4Br$ *[2]*

c i Tin and conc. HCl *[1]*

ii $C_6H_5NO_2 + 6[H] \rightarrow C_6H_5NH_2 + 2H_2O$ *[2]*

d i

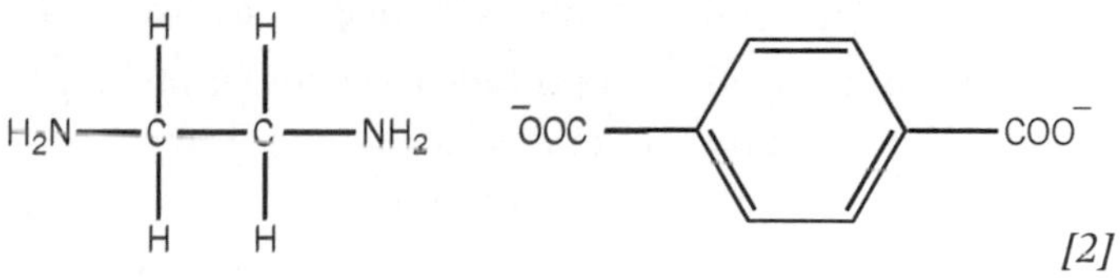

[2]

ii

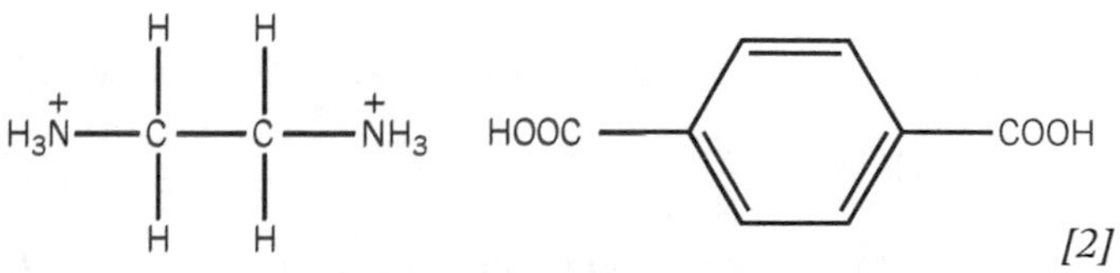

[2]

5 a i Non-superimposable mirror images *[1]*

ii

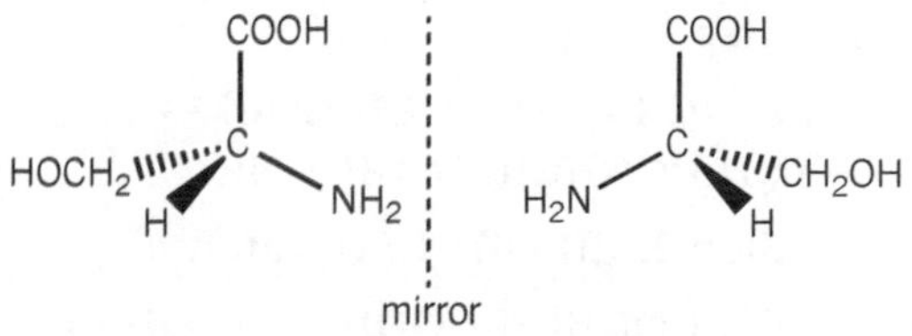

[2]

b

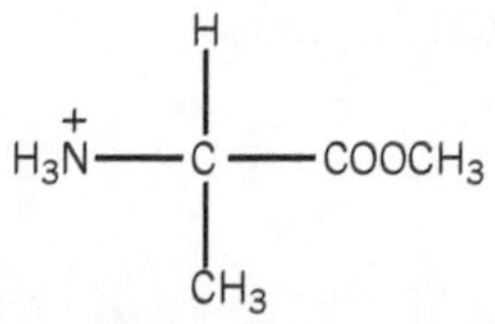

[2]

c 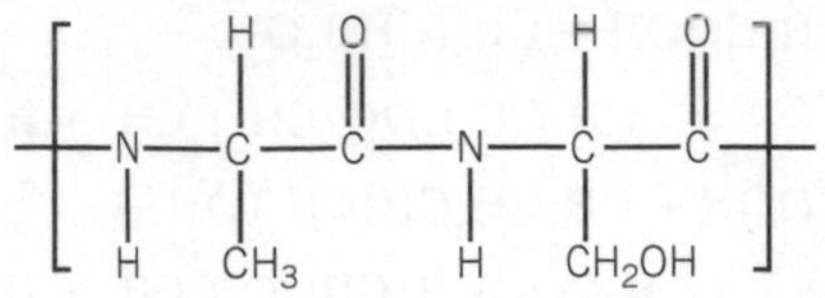

[2]

Chapter 28

1 a Aldehyde, ester, secondary amide. *[3]*

b Primary amide, phenol, secondary alcohol, ketone. *[4]*

2 a Step 1: KCN/ethanol *[1]*

$CH_3Br + KCN \rightarrow CH_3CN + KBr$ *[1]*

Step 2: H_2/Ni catalyst *[1]*

$CH_3CN + 2H_2 \rightarrow CH_3CH_2NH_2$ *[1]*

b Step 1: HCOCl, $AlCl_3$ halogen carrier *[1]*

$C_6H_6 + HCOCl \rightarrow C_6H_5CHO + HCl$ *[1]*

Step 2: $NaCN/H_2SO_4$ *[1]*

$C_6H_5CHO + HCN \rightarrow C_6H_5CH(OH)CN$ *[1]*

Step 3: HCl(aq) reflux *[1]*

$C_6H_5CH(OH)CN + HCl + 2H_2O \rightarrow C_6H_5CH(OH)COOH + NH_4Cl$ *[1]*

3 a $C_6H_5COOCH_3 + NaOH \rightarrow C_6H_5COONa + CH_3OH$ *[1]*

$C_6H_5COONa + HCl \rightarrow C_6H_5COOH + NaCl$ *[1]*

b i Reflux *[1]*

ii Pear-shaped/round bottom flask, condenser. *[2]*

c When pH indicator paper turns red. *[1]*

d Filter under reduced pressure. *[1]*

Recrystallise using minimum volume of hot solvent. *[1]*

Filter pure product under reduced pressure. *[1]*

e Impure organic compounds have lower melting points, and melt over a wider temperature range, than the pure compound. *[2]*

f $n(C_6H_5COOCH_3) = 5.28/136.0 = 0.0388\,mol$ *[1]*

= theoretical $n(C_6H_5COOH) = 0.0388\,mol$

Actual $n(C_6H_5COOH) = 3.76/122.0 = 0.0308\,mol$ *[1]*

$\% \text{ yield} = \frac{0.0308}{0.0388} \times 100 = 79.4\%$ *[1]*

4 a i Start: amine and carboxylic acid *[1]*

target: secondary amide and ester *[1]*

ii Step 1: CH_3COCl *[1]*

$H_2NCH_2COOH + CH_3COCl \rightarrow CH_3CONHCH_2COOH + HCl$ *[1]*

Step 2: CH_3OH/H_2SO_4 catalyst *[1]*

$CH_3CONHCH_2COOH + CH_3OH \rightarrow CH_3CONHCH_2COOCH_3 + H_2O$ *[1]*

b i Start: aldehyde *[1]*

target: alkene and carboxylic acid *[1]*

ii Step 1: $NaCN/H_2SO_4$ *[1]*

$CH_3CHO + HCN \rightarrow CH_3CH(OH)CN$ *[1]*

Step 2: HCl(aq) and reflux *[1]*

$CH_3CH(OH)CN + HCl + 2H_2O \rightarrow CH_3CH(OH)COOH + NH_4Cl$ *[1]*

Step 3: Conc H_2SO_4 catalyst *[1]*

$CH_3CH(OH)COOH \rightarrow H_2C{=}CHCOOH + H_2O$ *[1]*

Chapter 29

1 C *[1]*; **2** B *[1]*; **3** A *[1]*; **4** C *[1]*

5 C : H : O = 64.62/12.0 : 10.77/1.0 : 24.61/16.0 *[1]*

= 5.385 : 10.77 : 1.54 = $C_7H_{14}O_2$ *[1]*

From M^+ peak, $M_r = 130$ so molecular formula = $C_7H_{14}O_2$ *[1]*

IR Peak at $1720\,cm^{-1}$ indicates presence of C=O. *[1]*

Absence of peaks at $2500–3300\,cm^{-1}$ or $3200–3600\,cm^{-1}$ indicates no –OH group. *[1]*

^{1}NMR 3 peaks → 3 types of proton *[1]*

peak area ratio → 2 : 9 : 3 (from left)
→ $CH_2 : (CH_3)_3 : CH_3$ *[1]*

CH_2 quartet at 2.3 ppm
→ CH_2 adjacent to CH_3 and C=O: $CH_3CH_2C{=}O$

$(CH_3)_3$ singlet at 1.4 ppm
→ $(CH_3)_3$ adjacent to C with no protons: $(CH_3)_3C$

CH_3 triplet at 1.2 ppm
→ CH_3 adjacent to CH_2: CH_3CH_2 *[3]*

Compound **A** is the ester: $CH_3CH_2COOC(CH_3)_3$ *[1]*

Answers to synoptic questions

1

a i $\Delta S = [(4 \times 210.7) + (6 \times 188.7)] - [(4 \times 192.3) + (5 \times 205.0)] = +180.8\,\text{J mol}^{-1}\,\text{K}^{-1}$ *[2]*

$\Delta H = [(4 \times 90.2) + (6 \times -241.8)] - [(4 \times -46.1)] = -905.6\,\text{kJ mol}^{-1}$ *[2]*

$\Delta S = 180.8/1000 = 0.1808\,\text{kJ mol}^{-1}\,\text{K}^{-1}$

$\Delta G = -905.6 - 900 \times 0.1808 = -1068.32\,\text{kJ mol}^{-1}$ *[2]*

ii The forward reaction is exothermic. *[1]*

There are fewer gaseous moles of products. *[1]*

Decreasing temperature and increasing pressure shifts equilibrium position towards NO_2. *[1]*

iii $K_p = \dfrac{p(NO_2)^2}{p(NO)^2 \times p(O_2)}$ *[1]*

$p(NO) = \dfrac{1.80}{12.0} \times 10.5 = 1.575\,\text{atm}$ *[1]*

$p(O_2) = \dfrac{2.10}{12.0} \times 10.5 = 1.8375\,\text{atm}$ *[1]*

$p(NO_2) = \dfrac{8.10}{12.0} \times 10.5 = 7.0875\,\text{atm}$ *[1]*

$K_p = \dfrac{7.0875^2}{1.575^2 \times 1.8375} = 11.0\,\text{atm}^{-1}$ *[2]*

iv $2NO_2 + H_2O \rightarrow HNO_3 + HNO_2$ *[1]*

NO_2, N = +4 → HNO_3, N = +5 → oxidation *[1]*

NO_2, N = +4 → HNO_2, N = +3 → reduction *[1]*

b i $Mg + 2HNO_3 \rightarrow Mg(NO_3)_2 + H_2$ *[1]*

ii $Mg + 4HNO_3 \rightarrow Mg(NO_3)_2 + 2NO_2 + 2H_2O$ *[2]*

2

a i Sodium chlorate(V) *[1]*

ii $n(Na) : n(Cl) : n(O) = \dfrac{18.78}{23.0} : \dfrac{28.98}{35.5} : \dfrac{52.24}{16.0}$

= 0.8165 : 0.8163 : 3.265 *[1]*

= 1 : 1 : 4; Formula of **A** = $NaClO_4$ *[1]*

Reaction with $AgNO_3(aq)$ shows that compound **B** contains Cl^- ions. *[1]*

B = NaCl *[1]*

iii $4NaClO_3 \rightarrow 3NaClO_4 + NaCl$ *[1]*

b $ClO_3^- + 3SO_2 + 3H_2O \rightarrow Cl^- + 3SO_4^{2-} + 6H^+$ *[2]*

c i

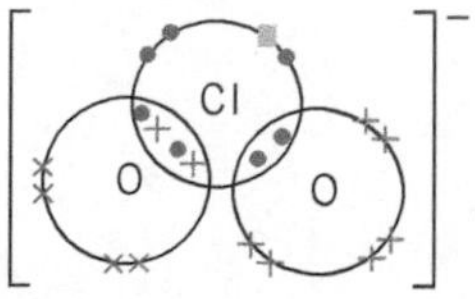

[2]

ii Shape: non-linear; Bond angle ~ 104.5° *[2]*

2 bonded regions (double and single bonds) and 2 lone pairs around central Cl atom. *[1]*

Lone pairs repel more than bonded pairs. *[1]*

d i $3NaClO \rightarrow NaClO_3 + 2NaCl$ *[1]*

ii

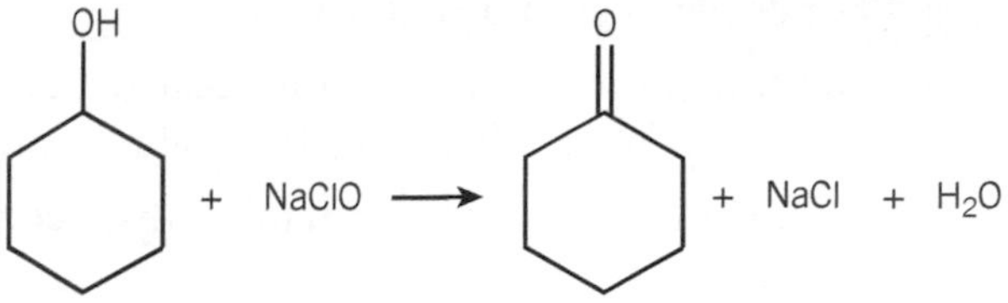

[2]

iii $HClO \rightleftharpoons H^+ + ClO^-$ *[1]*

$K_a = \dfrac{[H^+(aq)][ClO^-(aq)]}{[HClO(aq)]}$ *[1]*

$K_a = 10^{-7.53} = 2.95 \times 10^{-8}\,\text{mol dm}^{-3}$ *[1]*

$[H^+(aq)] = \sqrt{(2.95 \times 10^{-8} \times 0.250)}$
$= 8.59 \times 10^{-5}\,\text{mol dm}^{-3}$ *[1]*

$pH = -\log(8.59 \times 10^{-5}) = 4.07$ *[1]*

3

a Electrophilic substitution means that a benzene ring is present. *[1]*

Electrophilic addition means that an alkene/C=C is present. *[1]*

Compound **A** contains a benzene ring, C=C, and COOH → $C_9H_8O_2$. Molar mass = 148 = M^+ peak. *[1]*

Structure of **A** (the *E* stereoisomer):

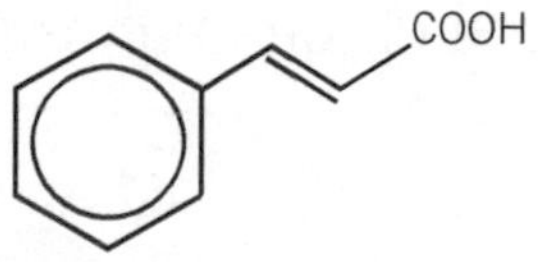

[1]

b i (2-)methylpropanedioic acid *[1]*

ii $C_4H_6O_4$, $M = 118\,\text{g mol}^{-1}$ *[2]*

iii C: 2 peaks
peak area ratio = 1 (COOH × 2) : 2 (CH_2 × 2) *[2]*

Singlet for (COOH × 2) and singlet for (CH_2 × 2) *[2]*

D: 3 peaks with peak area ratio
= 2 (COOH × 2) : 1 (CH) : 3 (CH_3) *[2]*

Singlet for (COOH × 2) *[1]*
quartet for CH adjacent to CH_3 *[1]*

doublet for CH_3 adjacent to CH *[1]*

iv $HOOCCH_2CH_2COOH + Na_2CO_3 \rightarrow NaOOCCH_2CH_2COONa + CO_2 + H_2O$ *[1]*

Neutralisation *[1]*

v **C** and **D** have 2 COOH groups which can each form H bonds between molecules. *[1]*

D is branched and there will be less contact between molecules than **C** giving weaker London forces. *[1]*

E has one COOH group which can form H bonds between molecules. *[1]*

F has no COOH groups and there are no H bonds between molecules. *[1]*

Order of strength of intermolecular forces = **C** > **D** > **E** > **F** which matches boiling points. *[1]*

c i Dissolve **G** in less than 250 cm^3 of distilled water in beaker. *[1]*

Transfer solution to a 250.0 cm^3 volumetric flask and transfer washings from beaker to flask. *[1]*

Make up to the mark with distilled water and invert flask several times to ensure mixing. *[1]*

ii $\% \text{ error} = \dfrac{0.05 \times 2}{24.75} \times 100 = 0.40\%$

(2 readings with ±0.05 error for each) *[1]*

iii $\text{Mean titre} = \dfrac{25.10 + 25.20}{2} = 25.15 \text{ cm}^3$

(Using only concordant titres) *[1]*

$n(\text{NaOH}) = 0.200 \times \dfrac{25.15}{1000} = 5.03 \times 10^{-3} \text{ mol}$ *[1]*

$n(\mathbf{G}) = \dfrac{5.03 \times 10^{-3}}{2} = 2.515 \times 10^{-3} \text{ mol}$ *[1]*

$n(\mathbf{G})$ in 250.0 cm^3 solution $= 10 \times 2.515 \times 10^{-3} = 2.515 \times 10^{-2} \text{ mol}$ *[1]*

$M(\mathbf{G}) = \dfrac{m}{n} = \dfrac{3.672}{2.515 \times 10^{-2}} = 146$ *[1]*

G = $HOOCCH_2CH_2CH_2CH_2COOH$ *[1]*

4

a $Ag^+(aq) + Cl^-(aq) \rightarrow AgCl(s)$ *[1]*

$\mathbf{M}Cl_2(aq) + 2AgNO_3(aq) \rightarrow \mathbf{M}(NO_3)_2(aq) + 2AgCl(s)$ *[1]*

b $n(\text{AgCl}) = \dfrac{1.150}{143.4} = 8.02 \times 10^{-3} \text{ mol}$ *[1]*

$n(\mathbf{M}Cl_2) = \dfrac{8.02 \times 10^{-3}}{2} = 4.01 \times 10^{-3}$ *[1]*

$M(\mathbf{M}Cl_2) = \dfrac{m}{n} = \dfrac{18.464 - 17.828}{4.01 \times 10^{-3}} = \dfrac{0.636}{4.01 \times 10^{-3}}$

$= 158.6 \text{ g mol}^{-1}$ *[1]*

A_r of **M** = 158.6 – 71.0 = 87.6;
M = Sr and anhydrous chloride = $SrCl_2$ *[1]*

c Mass of H_2O = 18.898 – 18.464 = 0.434 g;

$n(H_2O) = \dfrac{0.434}{18.0} = 0.0241 \text{ mol}$ *[1]*

$n(\mathbf{M}Cl_2) : n(H_2O) = 4.01 \times 10^{-3} : 0.0241 = 1 : 6$ *[1]*

Formula hydrated chloride = $SrCl_2{\bullet}6H_2O$ *[1]*

d Heat to constant mass (heat crucible again and reweigh; continue until mass is constant). *[1]*

e i When dried, and weighed, the precipitate would contain some $\mathbf{M}(NO_3)_2$. *[1]*

$n(\text{AgCl})$ and $n(\mathbf{M}Cl_2)$ would appear to be greater leading to a smaller value of $M(MCl_2)$ and a smaller value for A_r of the metal **M**. *[1]*

ii The mass of AgCl would be less than if all the Cl^- ions had reacted. *[1]*

$n(\text{AgCl})$ and $n(MCl_2)$ would appear to be smaller leading to a greater value of $M(MCl_2)$ and a greater value for A_r of the metal **M**. *[1]*

Periodic table

Key
atomic number
Symbol
name
relative atomic mass

(1)	(2)											(3)	(4)	(5)	(6)	(7)	(0)
1	2	3	4	5	6	7	8	9	10	11	12	13	14	15	16	17	18
1 **H** hydrogen 1.0																	2 **He** helium 4.0
3 **Li** lithium 6.9	4 **Be** beryllium 9.0											5 **B** boron 10.8	6 **C** carbon 12.0	7 **N** nitrogen 14.0	8 **O** oxygen 16.0	9 **F** fluorine 19.0	10 **Ne** neon 20.2
11 **Na** sodium 23.0	12 **Mg** magnesium 24.3											13 **Al** aluminium 27.0	14 **Si** silicon 28.1	15 **P** phosphorus 31.0	16 **S** sulfur 32.1	17 **Cl** chlorine 35.5	18 **Ar** argon 39.9
19 **K** potassium 39.1	20 **Ca** calcium 40.1	21 **Sc** scandium 45.0	22 **Ti** titanium 47.9	23 **V** vanadium 50.9	24 **Cr** chromium 52.0	25 **Mn** manganese 54.9	26 **Fe** iron 55.8	27 **Co** cobalt 58.9	28 **Ni** nickel 58.7	29 **Cu** copper 63.5	30 **Zn** zinc 65.4	31 **Ga** gallium 69.7	32 **Ge** germanium 72.6	33 **As** arsenic 74.9	34 **Se** selenium 79.0	35 **Br** bromine 79.9	36 **Kr** krypton 83.8
37 **Rb** rubidium 85.5	38 **Sr** strontium 87.6	39 **Y** yttrium 88.9	40 **Zr** zirconium 91.2	41 **Nb** niobium 92.9	42 **Mo** molybdenum 95.9	43 **Tc** technetium	44 **Ru** ruthenium 101.1	45 **Rh** rhodium 102.9	46 **Pd** palladium 106.4	47 **Ag** silver 107.9	48 **Cd** cadmium 112.4	49 **In** indium 114.8	50 **Sn** tin 118.7	51 **Sb** antimony 121.8	52 **Te** tellurium 127.6	53 **I** iodine 126.9	54 **Xe** xenon 131.3
55 **Cs** caesium 132.9	56 **Ba** barium 137.3	57–71 lanthanoids	72 **Hf** hafnium 178.5	73 **Ta** tantalum 180.9	74 **W** tungsten 183.8	75 **Re** rhenium 186.2	76 **Os** osmium 190.2	77 **Ir** iridium 192.2	78 **Pt** platinum 195.1	79 **Au** gold 197.0	80 **Hg** mercury 200.6	81 **Tl** thallium 204.4	82 **Pb** lead 207.2	83 **Bi** bismuth 209.0	84 **Po** polonium	85 **At** astatine	86 **Rn** radon
87 **Fr** francium	88 **Ra** radium	89–103 actinoids	104 **Rf** rutherfordium	105 **Db** dubnium	106 **Sg** seaborgium	107 **Bh** bohrium	108 **Hs** hassium	109 **Mt** meitnerium	110 **Ds** darmstadtium	111 **Rg** roentgenium	112 **Cn** copernicium		114 **Fl** flerovium		116 **Lv** livermorium		

57 **La** lanthanum 138.9	58 **Ce** cerium 140.1	59 **Pr** praseodymium 140.9	60 **Nd** neodymium 144.2	61 **Pm** promethium 144.9	62 **Sm** samarium 150.4	63 **Eu** europium 152.0	64 **Gd** gadolinium 157.2	65 **Tb** terbium 158.9	66 **Dy** dysprosium 162.5	67 **Ho** holmium 164.9	68 **Er** erbium 167.3	69 **Tm** thulium 168.9	70 **Yb** ytterbium 173.0	71 **Lu** lutetium 175.0
89 **Ac** actinium	90 **Th** thorium 232.0	91 **Pa** protactinium	92 **U** uranium 238.1	93 **Np** neptunium	94 **Pu** plutonium	95 **Am** americium	96 **Cm** curium	97 **Bk** berkelium	98 **Cf** californium	99 **Es** einsteinium	100 **Fm** fermium	101 **Md** mendelevium	102 **No** nobelium	103 **Lr** lawrencium

Data sheets

General Information

Molar gas volume = 24.0 $dm^3\ mol^{-1}$ at room temperature and pressure, RTP

Avogadro constant, $N_A = 6.02 \times 10^{23}\ mol^{-1}$

Specific heat capacity of water, $c = 4.18\ J\ g^{-1}\ K^{-1}$

Ionic product of water, $K_w = 1.00 \times 10^{-14}\ mol^2\ dm^{-6}$ at 298 K

1 tonne = 10^6 g

Arrhenius equation: $k = Ae^{-Ea/RT}$ or $\ln k = -E_a/RT + \ln A$

Gas constant, $R = 8.314\ J\ mol^{-1}\ K^{-1}$

Characteristic infrared absorptions in organic molecules

Bond	Location	Wavenumber / cm^{-1}
C–C	Alkanes, alkyl chains	750–1100
C–X	Haloalkanes (X = Cl, Br, I)	500–800
C–F	Fluoroalkanes	1000–1350
C–O	Alcohols, esters, carboxylic acids	1000–1300
C=C	Alkenes	1620–1680
C=O	Aldehydes, ketones, carboxylic acids, esters, amides, acyl chlorides and acid anhydrides	1630–1820
aromatic C=C	Arenes	Several peaks in range 1450–1650 (variable)
C≡N	Nitriles	2220–2260
C–H	Alkyl groups, alkenes, arenes	2850–3100
O–H	Carboxylic acids	2500–3300 (broad)
N–H	Amines, amides	3300–3500
O–H	Alcohols, phenols	3200–3600

^{13}C NMR chemical shifts relative to TMS

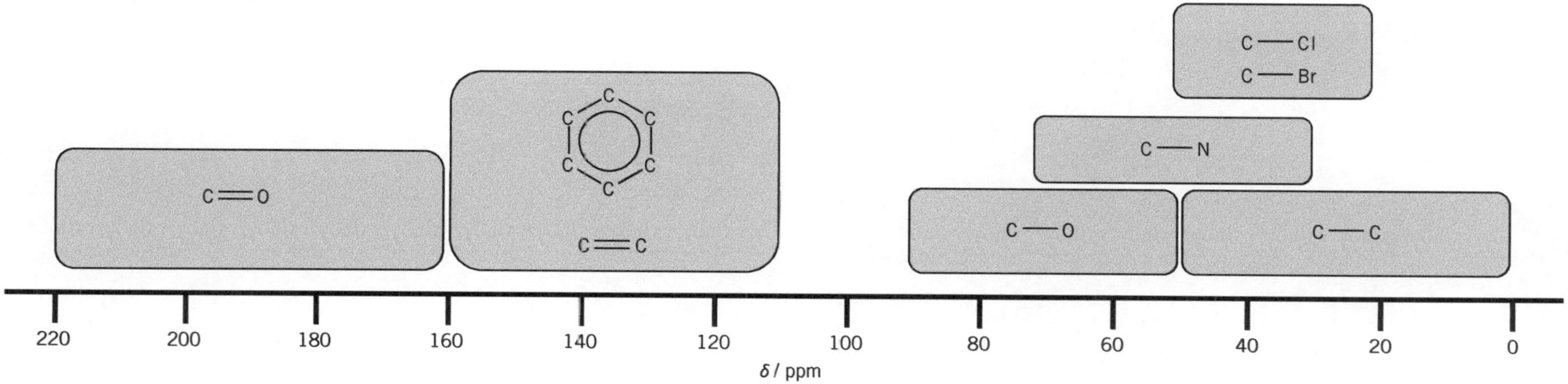

^{1}H NMR chemical shifts relative to TMS

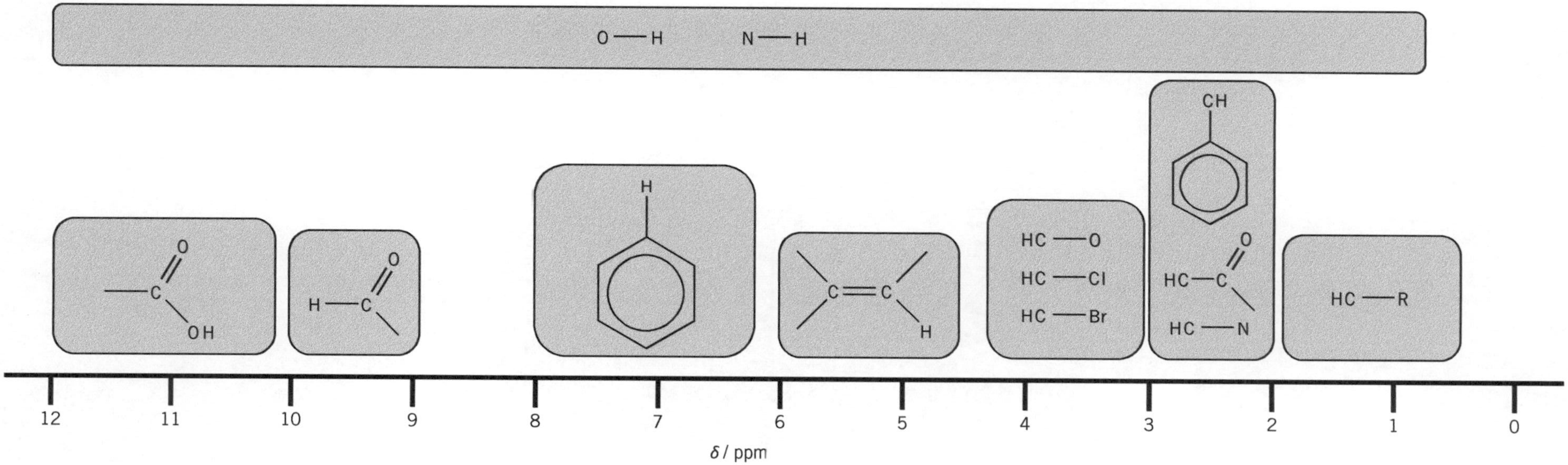

Chemical shifts are variable and can vary depending on the solvent, concentration and substituents. As a result, shifts may be outside the ranges indicated above.

OH and **NH** chemical shifts are very variable and are often broad. Signals are not usually seen as split peaks.

Note that **CH** bonded to 'shifting groups' on either side, e.g. O–CH_2–C=-O, may be shifted more than indicated above.